Who Says Elephants Can't Dance?

Who Says Elephants Can't Dance?

Inside IBM's
Historic
Turnaround

Louis V. Gerstner, Jr.

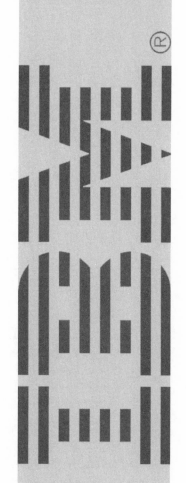

 HarperBusiness

An Imprint of HarperCollinsPublishers

HarperCollins books may be purchased for educational, business, or sales promotional use. For information, please write to: Special Markets Department, HarperCollins Publishers Inc., 10 East 53rd Street, New York, NY 10022.

Designed by William Ruoto

Library of Congress Cataloging-in-Publication Data
Gerstner, Louis V.
   Who says elephants can't dance? : inside IBM's historic turnaround / Louis V. Gerstner, Jr.
    p. cm.
   International Business Machines Corporation—Management. 2. International
Business Machines Corporation—History. 3. Computer industry—United States—
History. 4. Electronic office machines industry—United States—History.
5. Corporate turnarounds—United States—Case studies. I. Title.
HD9696.2.U64 I2545 2002
004.'068—dc21   2002027523

ISBN 0-06-052379-4 (alk. paper)

10 9 8 7 6 5 4 3 2 1

This book is dedicated to the thousands of IBMers
who never gave up on their company, their colleagues, and
themselves. They are the real heroes of the reinvention of IBM.

Contents

Foreword

I have never said to myself, "Gee, I think I want to write a book." I am not a book writer. Until now I haven't had the time or the inclination to lean back and reflect on my thirty-five years in business. I haven't had the patience it takes to sit down for a long time and create a book. Throughout my business life I have been wary of telling others how to manage their enterprises based on my personal experiences.

And, frankly, I wasn't sure if anyone would be interested in reading my thoughts. I read a lot of books, but not many about business. After a twelve-hour day at the office, who would want to go home and read about someone else's career at the office?

I have always believed you cannot run a successful enterprise from behind a desk. That's why, during my nine years as Chief Executive Officer of International Business Machines Corporation, I have flown more than 1 million miles and met with untold thousands of IBM customers, business partners, and employees. Over the past two years, after people began speculating that my retirement might be just around the corner, I thought I'd get a lot of big-picture questions that outgoing CEOs get about the economy, the world, and the future. Instead, I have been surprised by how many times—at big meetings and small ones, and even at private sessions with CEOs and heads of state—I was asked: "How did you save IBM?" "What was it like when you got there?" "What were the problems?" "What specific things did you do to bring the company back to life?" "What did you learn from the experience?"

They wanted to know, many of them said, because their own companies, organizations, or governments faced some of the same issues IBM had encountered during its very, very public near collapse in the early 1990s. Businesspeople outside the United States, confronted with the need to transform tradition-bound enterprises into tough and nimble players in a world economy, seemed particularly interested in the subject.

More recently, after I had announced my intention to retire, I was amused to read an editorial in an American newspaper, *USA Today,* that said, in effect, it hoped Gerstner was going to do something more useful than write a book and play golf. Nice thought, but since the announcement, I got thousands of letters and e-mails, and the most frequent sentiment was, again, that I should tell what I had learned from my tenure at IBM. (I was also invited to appear in a TV commercial with golf pros Jack Nicklaus and Gary Player.) You might say that I concluded, a little reluctantly, that the easiest way for me to fulfill all this "popular demand" was to write a book and to hold off on serious golf for the time being.

So, here I am, ready to tell you the story of the revival of IBM.

Of course, this book would never have appeared without the heroes among my IBM colleagues who helped me restore IBM to a position of leadership. In many respects this is their book as much as it is mine. There were many such leaders, but clearly among the most important were Dennie Welsh and Sam Palmisano, who built our services company; John Thompson, who created our Software Group; Abby Kohnstamm, who took a cacophony of confused messages and melded them into one of the most powerful brand statements in the world; Nick Donofrio, who was my translator from the world of high tech to the world of management; Jerry York, Rick Thoman, and John Joyce, three great financial executives who instilled a level of productivity, discipline, and probing analysis into a company that appeared to value these quite lightly when I arrived; Larry Ricciardi, my colleague of many years, who brought his intellect and counsel to many of our critical decisions; and, finally, Tom Bouchard, who as head of Human Resources stood tall and took the heat as we transformed the IBM culture.

There are many more. In fact, there are thousands of IBMers who answered the call, put their shoulders to the wheel, and performed magnificently as we undertook an exhausting—at times frightening, but always exhilarating—journey to restore this extraordinary company. To all of them, I dedicate this book.

I wrote this book without the aid of a coauthor or a ghostwriter (which is why it's a good bet this is going to be my last book; I had no idea it would be so hard to do). I am responsible for any mistakes or confusion the reader may endure. The views expressed are mine and are not necessarily those of the IBM Corporation or any other IBMer.

I did have a great deal of help from some longtime IBM colleagues. They include Jon Iwata, Mark Harris, and Mike Wing, who made substantial contributions. Michele Andrle managed production of every draft and redraft deftly and patiently—an unbelievable amount of stitching and restitching. I want to thank them and everyone else who helped me.

Introduction

This is not my autobiography. I can't think of anyone other than my children who might want to read that book (and I'm not 100 percent sure they would, either). However, in the spirit of trying to provide some contextual background for my views, what follows is a brief historical perspective.

I was born on March 1, 1942, in Mineola, New York—the county seat of Nassau County, Long Island.

My father started work as a milk-truck driver and ultimately became a dispatcher at the F&M Schaeffer Brewing Company. My mother was a secretary, sold real estate, and eventually became an administrator at a community college. Along with three brothers—one older, two younger—I lived in the same house in Mineola until I left for college, in 1959.

We were a warm, tightly knit, Catholic, middle-class family. Whatever I have done well in life has been a result of my parents' influence. My father was a very private man with a great love of learning and inner strength that needed no approbation or reinforcement from broader audiences. My mother was enormously disciplined, hard-working, and ambitious for all her children. She drove us toward excellence, accomplishment, and success.

Education was a high priority in the Gerstner household. My parents remortgaged their house every four years to pay for schooling. I attended public grade school, then Chaminade, a Catholic high school in Mineola. I graduated in 1959 and was almost on my way to

Notre Dame when Dartmouth College offered me a substantial scholarship. It was a major benefit for our family finances, so I packed off for Dartmouth in September 1959, without ever having set foot on its campus.

Four years later I graduated with a degree in engineering science. I immediately went to Harvard Business School for two years. (Back then one could leave undergraduate school and go directly to business school, a practice that has since, for the most part, been abandoned by business schools.)

Then, at the tender age of 23, I emerged from Harvard and went into business.

I joined the management consulting firm of McKinsey & Company in New York City in June 1965. My first assignment was to conduct an executive compensation study for the Socony Mobil Oil Co. I'll never forget my first day on that project. I knew nothing about executive compensation, and absolutely nothing about the oil industry. Thank goodness I was the low man on the totem pole, but in the McKinsey world one was expected to get up to speed in a hurry. Within days I was out meeting with senior executives decades older than I was.

Over the next nine years I advanced to the level of senior partner at McKinsey. I was responsible for its finance practice and was a member of its senior leadership committee. I was the partner in charge of three major clients, two of which were financial services companies.

The most important thing I learned at McKinsey was the detailed process of understanding the underpinnings of a company. McKinsey was obsessive about deep analysis of a company's marketplace, its competitive position, and its strategic direction.

When I reached my early thirties, it became clear to me that I didn't want to stay in consulting as a career. Although I enjoyed the intellectual challenge, the fast pace, and the interaction with top-ranking senior people, I found myself increasingly frustrated playing the role of an advisor to the decision makers. I remember saying to myself, "I no longer want to be the person who walks into the room and presents a report to a person sitting at the other end of the table; I want to be the person sitting in that chair—the one who makes the decisions and carries out the actions."

Like many other successful McKinsey partners, I had gotten a number of offers to join my clients over the years, but none of the proposals seemed attractive enough to make me want to leave. In 1977, however, I received and accepted an offer from American Express, which was my largest client at that time, to join it as the head of its Travel Related Services Group (basically, the American Express Card, Traveler's Checks, and Travel Office businesses). I stayed at American Express for eleven years, and it was a time of great fun and personal satisfaction. Our team grew Travel Related Services earnings at a compounded rate of 17 percent over a decade; expanded the number of cards issued from 8 million to nearly 31 million, and built whole new businesses around the Corporate Card, merchandise sales, and credit card processing industries.

I also learned a great deal. Early on I discovered, to my dismay, that the open exchange of ideas—in a sense, the free-for-all of problem solving in the absence of hierarchy that I had learned at McKinsey—doesn't work so easily in a large, hierarchical-based organization. I well remember stumbling in my first months when I reached out to people whom I considered knowledgeable on a subject regardless of whether they were two or three levels down from me in the organization. My team went into semi-revolt! Thus began a lifelong process of trying to build organizations that allow for hierarchy but at the same time bring people together for problem solving, regardless of where they are positioned within the organization.

It was also at American Express that I developed a sense of the strategic value of information technology. Think about what the American Express Card represents. It is a gigantic e-business, although we never thought about it in those terms in the 1970s. Millions of people travel the world with a sliver of plastic, charging goods and services in many countries. Every month they receive a single bill listing those transactions, all translated into a single currency. Concurrently merchants are paid around the globe for transactions completed by hundreds, if not thousands, of people whom they do not know and may never see again. All of this is done for the most part electronically, with massive data processing centers worldwide. The technology imperative of this business was something I wrestled with for many years.

This was also when I first discovered the "old IBM." I'll never forget the day one of my division managers called and said that he had recently installed an Amdahl computer in a large data center that had historically been 100-percent IBM equipped. He said that his IBM representative had arrived that morning and told him that IBM was withdrawing all support for his massive data processing center as a result of the Amdahl decision. I was flabbergasted. Given that American Express was at that time one of IBM's largest customers, I could not believe that a vendor had reacted with this degree of arrogance. I placed a call immediately to the office of the chief executive of IBM to ask if he knew about and condoned this behavior. I was unable to reach him and was shunted off to an AA (administrative assistant) who took my message and said he would pass it on. Cooler (or, should I say, smarter) heads prevailed at IBM and the incident passed. Nevertheless, it did not go out of my memory.

I left American Express on April 1, 1989, to accept what some in the media called at the time the "beauty contest" of the decade. RJR Nabisco, a huge packaged-goods company that had been formed a few years earlier through the merger of Nabisco and R. J. Reynolds Tobacco Company, was rated the ninth-most-admired company in America when the headhunters called me. The organization had just gone through one of the wildest adventures in modern American business history: an extraordinary bidding contest among various investment firms to take the company private through a leveraged buyout (LBO). The winning bid was made by the venture capital firm of Kohlberg Kravis Roberts & Co. (KKR). Soon afterward, KKR sought me out to become chief executive of the now private and heavily indebted company.

For the next four years I became immersed in a whole new set of challenges. While I understood well from my American Express days the ongoing demands of a consumer products company, I really spent most of my time at RJR Nabisco managing an extraordinarily complex and overburdened balance sheet. The LBO bubble of the 1980s burst shortly after the RJR Nabisco transaction, sending a tidal wave of trouble over this deal. In hindsight, KKR paid too much for the company, and the next four years became a race to refinance the balance sheet, while trying to keep some semblance of order in the many individual businesses of the company. It was a wild scene. We had to sell $11 billion worth of assets in the first twelve months. We had debt that paid interest rates as high as 21 percent a year. We had lender and creditor committees galore and, of course, the cleanup from the profligate spending of the prior management. (For example, when I arrived we had thirty-two professional athletes on our payroll—all part of "Team Nabisco.")

That was a difficult time for me. I love building businesses, not disassembling them. However, we all have an opportunity to learn in everything we do. I came away from this experience with a profound appreciation of the importance of cash in corporate performance— "free cash flow" as the single most important measure of corporate soundness and performance.

I also came away with a greater sense of the relationship between management and owners. I had experienced this at McKinsey, which was a private company owned by its partners. The importance of managers being aligned with shareholders—not through risk-free instruments like stock options, but through the process of putting their own money on the line through direct ownership of the company— became a critical part of the management philosophy I brought to IBM.

By 1992 it was clear to all that while RJR Nabisco itself was doing quite well, the LBO was not going to produce the financial returns the owners had expected. It was clear to me that KKR was headed for the exit, so it made sense for me to do the same. This book, which starts on the next page, picks up my story from there.

PART I

Grabbing Hold

The Courtship

On December 14, 1992, I had just returned from one of those always well-intentioned but rarely stimulating charity dinners that are part of a New York City CEO's life, including mine as CEO of RJR Nabisco. I had not been in my Fifth Avenue apartment more than five minutes when my phone rang with a call from the concierge desk downstairs. It was nearly 10 P.M. The concierge said, "Mr. Burke wants to see you as soon as possible this evening."

Startled at such a request so late at night in a building in which neighbors don't call neighbors, I asked which Mr. Burke, where is he now, and does he really want to see me face to face this evening?

The answers were: "Jim Burke. He lives upstairs in the building. And, yes, he wants very much to speak to you tonight."

I didn't know Jim Burke well, but I greatly admired his leadership at Johnson & Johnson, as well as at Partnership for a Drug-Free America. His handling of the Tylenol poisoning crisis years earlier had made him a business legend. I had no idea why he wanted to see me so urgently. When I called, he said he would come right down.

When he arrived he got straight to the point: "I've heard that you may go back to American Express as CEO, and I don't want you to do that because I may have a much bigger challenge for you." The reference to American Express was probably prompted by rumors that I was going to return to the company where I had worked for eleven years. In fact, in mid-November 1992, three members of the American Express board had met secretly with me at the Sky Club in New York City to ask that I come back. It's hard to say if I was surprised—Wall Street and the media were humming with speculation that then CEO Jim Robinson was under board pressure to step down. However, I told the three directors politely that I had no interest in returning to American Express. I had loved my tenure there, but I was not going back to fix mistakes I had fought so hard to avoid. (Robinson left two months later.)

I told Burke I wasn't returning to American Express. He told me that the top position at IBM might soon be open and he wanted me to consider taking the job. Needless to say, I was very surprised. While it was widely known and reported in the media that IBM was having serious problems, there had been no public signs of an impending change in CEOs. I told Burke that, given my lack of technical background, I couldn't conceive of running IBM. He said, "I'm glad you're not going back to American Express. And please, keep an open mind on IBM." That was it. He went back upstairs, and I went to bed thinking about our conversation.

The media drumbeat intensified in the following weeks. *Business-Week* ran a story titled "IBM's Board Should Clean Out the Corner Office." *Fortune* published a story, "King John [Akers, the chairman and CEO] Wears an Uneasy Crown." It seemed that everyone had advice about what to do at IBM, and reading it, I was glad I wasn't there. The media, at least, appeared convinced that IBM's time had long passed.

The Search

On January 26, 1993, IBM announced that John Akers had decided to retire and that a search committee had been formed to consider outside and internal candidates. The committee was headed by Jim Burke. It didn't take long for him to call.

I gave Jim the same answer in January as I had in December: I wasn't qualified and I wasn't interested. He urged me, again: "Keep an open mind."

He and his committee then embarked on a rather public sweep of the top CEOs in America. Names like Jack Welch of General Electric, Larry Bossidy of Allied Signal, George Fisher of Motorola, and even Bill Gates of Microsoft surfaced fairly quickly in the press. So did the names of several IBM executives. The search committee also conducted a series of meetings with the heads of many technology companies, presumably seeking advice on who should lead their number one competitor! (Scott McNealy, CEO of Sun Microsystems, candidly told one reporter that IBM should hire "someone lousy.") In what was believed to be a first-of-its-kind transaction, the search committee hired two recruiting firms in order to get the services of the two leading recruiters—Tom Neff of Spencer Stuart Management Consultants N.V., and Gerry Roche of Heidrick & Struggles International, Inc.

In February I met with Burke and his fellow search committee member, Tom Murphy, then CEO of Cap Cities/ABC. Jim made an emphatic, even passionate pitch that the board was not looking for a technologist, but rather a broad-based leader and change agent. In fact, Burke's message was consistent throughout the whole process. At the time the search committee was established, he said, "The committee members and I are totally open-minded about who the new person will be and where he or she will come from. What is critically important is the person must be a proven, effective leader—one who is skilled at generating and managing change."

Once again, I told Burke and Murphy that I really did not feel qualified for the position and that I did not want to proceed any further with the process. The discussion ended amicably and they went off, I presumed, to continue the wide sweep they were carrying out, simultaneously, with multiple candidates.

What the Experts Had to Say

I read what the press, Wall Street, and the Silicon Valley computer visionaries and pundits were saying about IBM at that time. All of it certainly fueled my skepticism and, I believe, that of many of the other candidates.

Most prominent were two guys who seemed to pop up everywhere you looked, in print and on TV—Charles Morris and Charles Ferguson. They had written a book, *Computer Wars,* that took a grim view of IBM's prospects. They stated: "There is a serious possibility that IBM is finished as a force in the industry. Bill Gates, the software tycoon whom everybody in the industry loves to hate, denies having said in an unguarded moment that IBM 'will fold in seven years.' But Gates may be right. IBM is now an also-ran in almost every major computer technology introduced since 1980. . . . Traditional big computers are not going to disappear overnight, but they are old technology, and the realm in which they hold sway is steadily shrinking. The brontosaurus moved deeper into the swamps when the mammals took over the forests, but one day it ran out of swamps."

Their book concluded that "the question for the present is whether IBM can survive. From our analysis thus far, it is clear that we think its prospects are very bleak."

Morris and Ferguson wrote a longer, more technical, and even grimmer report on IBM and sold it to corporations and institutions for a few thousand dollars per copy. Among others, it frightened a number of commercial banks that were lenders to IBM.

Paul Carroll, IBM's beat reporter at *The Wall Street Journal,* published a book that year chronicling IBM's descent. In it, he said: "The world will look very different by the time IBM pulls itself together—

assuming it *can* pull itself together—and IBM will never again hold sway over the computer industry."

Even *The Economist*—understated and reliable—over the span of six weeks, published three major stories and one lengthy editorial on IBM's problems. "Two questions still hang over the company," its editors wrote. "In an industry driven by rapid technological change and swarming with smaller, nimbler firms, can a company of IBM's size, however organized, react quickly enough to compete? And can IBM earn enough from expanding market segments such as computer services, software, and consulting to offset the horrifying decline in mainframe sales, from which it has always made most of its money?

"The answer to both questions may be no."

And, said the usually sober *Economist,* "IBM's humiliation is already being viewed by some as a defeat for America."

The Decision

The turning point in my thinking occurred over Presidents' Day weekend in February 1993. I was at my house in Florida, where I love to walk the beach, clearing and settling my mind. It's very therapeutic for me. During an hour's walk each day that weekend, I realized that I had to think differently about the IBM situation. What prompted my change of heart was what was happening at RJR Nabisco. As I noted in the Introduction, it had become clear that KKR had given up on making its leveraged buyout work as planned. There were two reasons for this. First, as discussed in Bryan Burroughs and John Helyar's book *Barbarians at the Gate,* in the fury and madness of the bidding process in 1988, KKR overpaid for RJR Nabisco. This meant that despite achieving all of the restructuring objectives of the LBO, there simply wasn't enough operating leverage to produce the projected returns. Second, the operating returns from the tobacco business were under pressure as a result of a price war started by Philip Morris soon after the RJR Nabisco buy out. Philip Morris was simply following the advice of Ray Kroc, founder of McDonald's, who'd once said, "When you see your competitor drowning, grab a fire hose and put it in his mouth."

KKR obviously was working on an exit strategy. As I walked the beach that February, I decided I should be doing the same thing. And so, as much as anything else, the view that I would not be at RJR Nabisco too much longer was what got me thinking more about the IBM proposal.

I called Vernon Jordan, the Washington attorney who was a longtime friend as well as a director of RJR Nabisco, and asked his advice. He confirmed my feeling that KKR was ready to move out of RJR Nabisco and that this phase of the company's tumultuous life was coming to an end. Also, it was clear that Jim Burke had already talked with Vernon, because Vernon knew I was on the IBM list. His advice was, as usual, to the point. He said, "IBM is the job you have been in training for since you left Harvard Business School. Go for it!"

I suppose there was a second reason I changed my mind. I have always been drawn to a challenge. The IBM proposition was daunting and almost frightening, but it was also intriguing. The same was true of RJR Nabisco when I'd joined it in 1989. I think it is fair to say that from February 15 on, I was prepared to consider taking on IBM and its problems. Vernon got word to Jim Burke that I might be in play after all. I began to organize my questions and concerns for Burke and his committee.

When Burke called later that week, I told him that I would take a look at the IBM job. I told him I would need a lot more information, particularly about the company's short- and intermediate-term prospects. The dire predictions of the media and the pundits had me worried. I had learned a hard lesson at RJR Nabisco: A company facing too many challenges can run out of cash very quickly.

I told Burke that I wanted to meet with Paul Rizzo. Paul had been an executive at IBM in the 1980s. I had met him on several occasions and admired him greatly. He had retired from IBM in 1987 but had been called back by the IBM Board of Directors in December 1992 to work with John Akers to stem the decline of the company. I told Burke during that February phone call that I wanted to go over the budget and plans for 1993 and 1994 with Rizzo.

Jim moved quickly, and on February 24, at the Park Hyatt hotel in Washington, D.C., where I was attending a meeting of The Business Council, I broke away for an hour and a half to meet secretly with Paul in my hotel room. He had brought me the current financials and budgets for the company.

The discussion that ensued was very sobering. IBM's sales and profits were declining at an alarming rate. More important, its cash position was getting scary. We went over each product line. A lot of the information was difficult to evaluate. However, Paul clearly underscored the make-or-break issue for the company: He said that mainframe revenue had dropped from $13 billion in 1990 to a projection of less than $7 billion in 1993, and if it did not level off in the next year or so, all would be lost. He also confirmed that the reports in the press about IBM pursuing a strategy of breaking up the company into independent operating units was true. I thanked Paul for his honesty and insight and promised to treat the material with total confidentiality.

When he left the room, I was convinced that, on the basis of those documents, the odds were no better than one in five that IBM could be saved and that I should never take the position. A consumer products company has long-term brands that go on forever. However, that was clearly not the case in a technology company in the 1990s. There, a product could be born, rise, succeed wildly, crash, disappear, and be forgotten all within a few years. When I woke up the next morning, I was convinced IBM was not in my future. The company was slipping rapidly, and whether that decline could be arrested in time—by anyone—was at issue.

Still, Jim Burke would not give up. His persistence may have had more to do with a growing desperation to get anyone to take this job than it did with a particular conviction that I was the right candidate. I wondered at this point if he was just trying to keep a warm body in play.

Two weeks later I was back in Florida for a brief vacation. Burke and Murphy insisted on a meeting to pursue the issue one last time. We met in a new house that headhunter Gerry Roche and his wife had just built in a community near my own. Roche only played the role of host. In his new living room, it was Burke, Murphy, and me alone. I remember that it was a long afternoon.

Burke introduced the most novel recruitment argument I have ever heard: "You owe it to America to take the job." He said IBM was such a national treasure that it was my obligation to fix it.

I responded that what he said might be true only if I felt confident I could do it. However, I remained convinced the job was not doable—at least not by me.

Burke persisted. He said he was going to have President Bill Clinton call me and tell me that I had to take the job.

Tom Murphy, who tended to let Burke do most of the talking in our previous meetings, spoke up more frequently this time. Murph, as he is called by his friends, was quite persuasive in arguing that my track record as a change agent (his term) was exactly what IBM needed and that he believed there was a reasonable chance that, with the right leadership, the company could be saved. He reiterated what I'd heard from Burke, and even Paul Rizzo. The company didn't lack for smart, talented people. Its problems weren't fundamentally technical in nature. It had file drawers full of winning strategies. Yet, the company was frozen in place. What it needed was someone to grab hold of it and shake it back into action. The point Murphy came back to again and again was that the challenge for the next leader would begin with driving the kind of strategic and cultural change that had characterized a lot of what I'd done at American Express and RJR.

At the end of that long afternoon, I was prepared to make the most important career decision of my life. I said yes. In retrospect, it's almost hard for me to remember why. I suppose it was some of Jim Burke's patriotism and some of Tom Murphy's arguments playing to my gluttony for world-class challenges. At any rate, we shook hands and agreed to work out a financial package and announcement.

In hindsight, it's interesting that both Burke and Murphy were operating under the assumption provided by the management of IBM that a strategy of breaking up the company into independent units was the right one to pursue. What would they have said if they realized that not only was the company in financial trouble and had lost touch with its customers, but that it was also barreling toward a strategy of disaster?

I drove home that afternoon and told my family of my decision. As usual in my wonderful family, I got a mixed reaction. One of my children said, "Yes, go for it, Dad!" The other, more conservative child thought I had lost my mind. My wife, who had been quite wary of the idea originally, supported my decision and was excited about it.

[2]

The Announcement

Over the next ten days we worked out an employment contract. Doing so was not easy, for several reasons. The big one was the fact that RJR Nabisco was an LBO, and in an LBO the CEO is expected to align himself or herself with the owners and have a large equity position in his or her company. Consequently, at RJR Nabisco I owned 2.4 million shares and had options for 2.6 million more. In IBM, stock ownership was a *de minimus* part of executive compensation. The IBM board and human resources (HR) bureaucracy apparently did not share the view that managers should have a significant stake in the company. This was my first taste of the extraordinary insularity of IBM.

Somehow we worked everything out, and my next task was to tell KKR and the RJR board of my decision. That weekend, March 20–21, was the annual Nabisco Dinah Shore Golf Tournament. Nabisco invited all of its major customers to the event, and it was important for me to attend. I also knew that Henry Kravis, one of KKR's senior partners, would be there, and I decided that I would discuss my decision with him then. My name had already surfaced in the media as a candidate for the IBM job, and I knew that KKR and the RJR board were nervous. In meetings with KKR over the preceding weeks, there was a noticeable tension in the air. So, on Sunday, March 21, in my hotel room at the Dinah Shore tournament, I told Henry Kravis I was going to accept the IBM job. He was not happy, but true to form, he was polite and calm. He tried to talk me out of my decision, but I was firm that there was no going back. While we never discussed it, what was implicit in that conversation was the knowledge we shared that both of us were working on our exits from RJR. I just happened to finish sooner. (KKR started its exit a year later.)

The next day, Monday, I returned from California for the beginning of a very eventful week. The IBM board was meeting a week later. It became obvious that the search committee had begun to shut down its operations—because, one by one, the other rumored candidates for the IBM job started to announce or to leak to the media that they were not interested in the job. On Wednesday *The Wall Street Journal* reported that I was the only candidate; the next day, so did every other major business publication. It was time to get this whole ordeal over with. Burke and I agreed to announce on that Friday, March 26. That set off a mad scramble to organize both internal and external messages in the midst of a firestorm of leaks and headlines.

IBM made the announcement on Friday morning (even though the cover story of *BusinessWeek,* published that morning, already stated I had taken the job). A press conference began at 9:30 A.M. at the Hilton hotel in New York City. John Akers, Jim Burke, and I spoke. Burke wanted to explain the search process that had seemed so public and disjointed for three months. He made these comments during his opening statement: "There was only a handful of people in the world who were capable of handling this job. I want you to know that Lou Gerstner was on that original list, but we then did a worldwide search of well over one hundred twenty-five names, which we processed and kept reducing . . . and pretty well got back to the list that we started with. We gave those people on the list code names in an attempt to keep it out of the press—a vain attempt, I might add. You might be interested that Lou Gerstner was the first person I talked to on that list and consequently had the code name 'Able.' I knew all the other candidates—and I knew them all well. There isn't another candidate that could do this job any better than Lou Gerstner will. We made one specific offer for this job and only one, and that was to Louis Gerstner. While many people felt that technology was the key to this, there's a list of the specifications that we as a committee put together from the beginning. The fact is, there are fifteen things on the list and only one statement of the fifteen: 'Information and high-technology industry experience [are] highly desirable, but not opposed to considering extraordinary business leaders.' All of the others list qualities which are inherent in Lou Gerstner."

I knew my life was changing forever when I walked to the podium and three dozen photographers surged forward, and I had to conduct an entire press conference through nonstop, blinding camera flashes. As visible as American Express and RJR had been, this was something altogether different. I was now a public figure. This wasn't just any company—even any very big company. IBM was an institution—a global one—and its every move was scrutinized by the outside world. I was taking on a daunting challenge, and I'd be doing it in a fishbowl.

I am by nature a private person and, to be frank, I don't enjoy dealing with the press. On top of that, I looked around the industry, and as far as the eye could see there were (and still are) senior executives seeking the highest personal profile they could manage. I felt then, and I feel today, that while that kind of relentless publicity seeking generates a lot of coverage, and may even help the company in the short run, in the long run it damages corporate reputation and customer trust.

So I faced the cameras and lights that morning with mixed emotions. I was as full of adrenaline as I had ever been in my life. At the same time, I knew this was the big show and there was no escaping it.

My own remarks were brief. I was just trying to get through the ordeal without dealing with a lot of specific questions about why I felt I was qualified for the job and what I was going to do to fix IBM. But those were exactly the questions I got in a lengthy Q&A following the formal remarks. Needless to say, I provided little nourishment for the reporters. I simply had no idea what I would find when I actually arrived at IBM.

Meeting the IBM Team

After the press conference came a series of internal IBM meetings. As I look back at my schedule, I see that the first meeting the IBM human resources people had set up was a telephone conference call with the general managers of all the country operations around the world, underscoring that the power base of the company was the country leaders.

We then raced from Manhattan by helicopter some thirty miles north to the company's worldwide headquarters, in Armonk, New York. While I had been in some IBM facilities before as a customer, I had never been in the headquarters building. I will never forget my first impression. It reminded me of a government office—long, quiet corridor after long, quiet corridor of closed offices (quiet that was broken only by the presence of the almost blindingly bright orange carpeting). There was not a single indication in the artwork or other displays that this was a computer company. There was no computer in the CEO's office.

I was ushered into a large conference room to meet with the Corporate Management Board—roughly the top fifty people in the company. I don't remember what the women wore, but it was very obvious that all the men in the room were wearing white shirts, except me. Mine was blue, a major departure for an IBM executive! (Weeks later at a meeting of the same group, I showed up in a white shirt and found everyone else wearing other colors.)

When John Akers had suggested this meeting earlier in the week, he had assumed it was going to be simply an opportunity for me to meet the senior members assembled. However, I viewed it as a critical opportunity to introduce myself and, at least, set an initial agenda for my new colleagues. I worked hard in advance organizing what I wanted to say to the group. (In fact, in researching this book, I discovered detailed notes I had prepared—something I don't do very often for informal meetings.)

After John introduced me, the group sat politely, expecting nothing more than a "welcome and I'm delighted to be part of the team" salutation. Instead, I spoke for forty or forty-five minutes.

I started out explaining why I took the job—that I hadn't been looking for it, but had been asked to take on a responsibility that was important to our country's competitiveness and our economy's health. I didn't say it at the time, but it was my feeling that if IBM failed, there would be repercussions beyond the demise of one company. I indicated that I had no preconceived notions of what needed to be done and, from what I could tell, neither did the board. I said that for each of them (and for me!) there would be no special protection for past successes. But I clearly needed their help.

I then dealt with what I described as my early expectations: "If IBM is as bureaucratic as people say, let's eliminate bureaucracy fast. Let's decentralize decision making wherever possible, but this is not always the right approach; we must balance decentralized decision making with central strategy and common customer focus. If we have too many people, let's right-size fast; let's get it done by the end of the third quarter." I explained that what I meant by right-size is straightforward: "We have to benchmark our costs versus our competitors and then achieve best-in-class status." I also remarked that we had to stop saying that IBM didn't lay off people. "Our employees must find it duplicitous and out of touch with what has been going on for the last year." (In fact, since 1990, nearly 120,000 IBM employees had left the company, some voluntarily and some involuntarily, but the company had continued to cling to the fiction of "no layoffs.")

Perhaps the most important comments I made at that meeting regarded structure and strategy. At the time, the pundits and IBM's own leadership were saying that IBM should break itself up into smaller, independent units. I said, "Maybe that is the right thing to do, but maybe not. We certainly want decentralized, market-driven decision making. But is there not some unique strength in our ability to offer comprehensive solutions, a continuum of support? Can't we do that and also sell individual products?" (In hindsight it was clear that, even before I started, I was skeptical about the strategy of atomizing the company.)

I then talked about morale. "It is not helpful to feel sorry for ourselves. I'm sure our employees don't need any rah-rah speeches. We need leadership and a sense of direction and momentum, not just from me but from all of us. I don't want to see a lot of prophets of doom around here. I want can-do people looking for short-term victories and long-term excitement." I told them there was no time to focus on who created our problems. I had no interest in that. "We have little time to spend on problem definition. We must focus our efforts on solutions and actions."

Regarding their own career prospects, I noted that the press was saying that "the new CEO has to bring a lot of people in from the outside." I pointed out that I hoped this would not be the case, that IBM had always had a rich talent pool—perhaps the best in the world. I said, "If necessary, I will bring in outsiders, but you will each first get a chance to prove yourself, and I hope you will give me some time to prove myself to you. Everyone starts with a clean slate. Neither your successes nor failures in the past count with me."

I went on to summarize my management philosophy and practice:

- I manage by principle, not procedure.
- The marketplace dictates everything we should do.
- I'm a big believer in quality, strong competitive strategies and plans, teamwork, payoff for performance, and ethical responsibility.
- I look for people who work to solve problems and help colleagues. I sack politicians.
- I am heavily involved in strategy; the rest is yours to implement. Just keep me informed in an informal way. Don't hide bad information—I hate surprises. Don't try to blow things by me. Solve problems laterally; don't keep bringing them up the line.
- Move fast. If we make mistakes, let them be because we are too fast rather than too slow.
- Hierarchy means very little to me. Let's put together in meetings the people who can help solve a problem, regardless of position. Reduce committees and meetings to a minimum. No committee decision making. Let's have lots of candid, straightforward communications.
- I don't completely understand the technology. I'll need to learn it, but don't expect me to master it. The unit leaders must be the translators into business terms for me.

I then proposed that, based on my reading, we had five ninety-day priorities:

- Stop hemorrhaging cash. We were precariously close to running out of money.
- Make sure we would be profitable in 1994 to send a message to the world—and to the IBM workforce—that we had stabilized the company.
- Develop and implement a key customer strategy for 1993 and 1994—one that would convince customers that we were back

serving their interests, not just pushing "iron" (mainframes) down their throats to ease our short-term financial pressures.

- Finish right-sizing by the beginning of the third quarter.
- Develop an intermediate-term business strategy.

Finally, I laid out an assignment for the next thirty days. I asked for a ten-page report from each business unit leader covering customer needs, product line, competitive analysis, technical outlook, economics, both long- and short-term key issues, and the 1993–94 outlook.

I also asked all attendees to describe for me their view of IBM in total: What short-term steps could we take to get aggressive on customer relationships, sales, and competitive attacks? What should we be thinking about in our long-term and short-term business strategies?

In the meantime, I told everyone to go out and manage the company and not to talk to the press about our problems, and help me establish a travel schedule that would take me to customers and employees very early. "Let me know the meetings you are scheduled to hold over the next few weeks and recommend whether I should attend or not."

I asked if there were any questions. There were none. I walked around and shook everyone's hand and the meeting ended.

As I look back from the vantage point of nine years' tenure at IBM, I'm surprised at how accurate my comments proved to be. Whether it was the thoroughness of the press coverage, my experience as a customer, or my own leadership principles, what needed to be done—and what we did—was nearly all there in that forty-five-minute meeting four days before I started my IBM career.

The Official Election

On Tuesday of the following week, March 30, 1993, I attended the regularly scheduled IBM board meeting. It was at this gathering that I was elected Chairman of the Board and Chief Executive Officer, effective to begin two days later.

I walked into the meeting with a certain degree of trepidation. Jim Burke had told me a week earlier that there were two board members who were not totally happy with my selection as the new CEO. As I walked around the room shaking hands and greeting each of the seventeen directors in attendance (one was missing), I couldn't help but wonder who the two doubters were.

There are several things I remember well from that first meeting. The first was that there was an executive committee. Three of the eight members were current or former employees. I was taken by the fact that this board-within-a-board discussed in more detail the financial outlook for the company than the subsequent discussion held with the full Board of Directors.

The full board meeting focused on a wide range of subjects. It seemed to me from the agenda that it was a business-as-usual board meeting. There was a presentation from the Storage Division, which was being renamed AdStar as part of the overall corporate strategy to spin off the operating units. There were reports on business from the heads of domestic and international sales, discussion of a regulatory filing, and the approval of a proposed $440 million acquisition. If the directors felt there was a crisis, they were politely hiding it from me.

The meeting got a bit more animated during a report on financial affairs. Among the items reported was that the March quarter's gross margin on hardware had declined nineteen points from the prior year and that System/390 mainframe prices had declined 58 percent over the same period. The projection was for a loss of 50 cents a share for the quarter ending the next day. The cash situation was deteriorating fast. A major item of business was to approve a new financing plan authorizing the company to increase committed lines of bank credit to $4.7 billion and to raise $3 billion through the issuance of preferred stock and/or debt and securitization of United States trade receivables (selling, at a discount, "IOUs" from customers in order to get cash sooner).

It was clear there was a high degree of uncertainty surrounding the financial projections. The meeting ended. There were polite statements of "good luck" and "glad you're here," and everyone left.

John Akers and I then met to talk about the company. John and I had served together on the New York Times Company Board of Directors for several years, saw each other frequently at CEO-level events, and had a solid personal relationship prior to his departure from IBM. We were as comfortable as two people could be under the circumstances. We talked mostly about people. He was surprisingly candid about and critical of many of his direct reports. In reviewing my notes from the meeting, I guess I subsequently agreed with 75 percent of his appraisals. What struck me was why he could be so critical but still keep some of these executives in place. He had two favorites. One turned out to be one of my own. The other I let go before a year had passed.

Regarding business issues, John was preoccupied that day with IBM's microelectronics business. I learned that the company was deep into discussions with Motorola to form a joint venture and, in so doing, secure a partial exit from what John called the "technology business." I asked how imminent the decision was, and he said "very." Somewhat related to the Motorola deal was a proposal to license manufacturing rights for Intel microprocessors.

He said the basic research unit was not affordable and needed to be downsized. He was quite concerned about IBM's software business, mainframe business, and midrange products. As I look back at my notes, it is clear he understood most, if not all, of the business issues we tackled over the ensuing years. What's striking from my notes is the absence of any mention of culture, teamwork, customers, or leadership—the elements that turned out to be the toughest challenges at IBM.

John moved that day to an office in Stamford, Connecticut, and as far as I know, to his credit, he never looked back.

I went home with a deepening sense of fear. Could I pull this off? Who was going to help me?

Drinking from
a Fire Hose

On April 1, 1993, I began my IBM career (perhaps appropriately, April Fools' Day). IBM's stock stood at $13.[1] An op-ed piece in *The New York Times* greeted me with yet more advice on how to fix the company: "IBM has plenty of brains and button-downs. What it needs is bravado."

An IBM company car picked me up at my home in Connecticut at 6:45 A.M. and drove me not to the headquarters building, in Armonk, but to another of the many office complexes IBM owned at the time in Westchester County, New York. Consistent with my message to the senior management team the previous week, Ned Lautenbach, who then headed all of sales outside the United States (what IBM called "World Trade"), invited me to a meeting of all the country general managers that happened to be scheduled for that morning.

When I arrived at this large and spacious office building (it is now the headquarters of MasterCard International), I walked up to the front door and it was locked. A card reader was next to the door,

---

1   Adjusted for subsequent stock splits.

but I had not yet been issued a badge by IBM security. There I was, the new CEO, knocking helplessly on the door, hoping to draw someone's attention to let me in. After a while a cleaning woman arrived, checked me out rather skeptically, then opened the door—I suspect more to stop my pounding on the door than from any sense on her part that I belonged on the inside rather than the outside of the building.

I wandered around and eventually found the conference room where the meeting was just about to begin. I'll never forget my first impression of an IBM meeting. Arrayed around a long conference room were all the nobles of IBM's offshore, geographical fiefdoms. Behind them was a double row filled with younger executives. All the principals were white males, but the younger support staff was far more diverse. The meeting was an operations review, and each of the executives commented on his business. I noticed the backbenchers were scribbling furiously and occasionally delivered notes to the people at the table. It looked like a United States congressional hearing.

During a coffee break, I asked Ned Lautenbach, "Who are all these people who are clearly watching but not participating?"

He said, "Those are the executives' AAs."

And so it was at my first meeting on my first day at IBM that I encountered its solidly entrenched and highly revered administrative assistant program. Hundreds, if not thousands, of IBM middle- and senior-level executives had assistants assigned to them, drawn from the ranks of the best and brightest of the up-and-coming managers. The tasks were varied, but from what I could understand, AAs had primarily administrative duties and even, at times, secretarial chores. For the most part, AAs organized things, took notes, watched, and, hopefully, learned. What they didn't do was interact with customers, learn the guts of the business, or develop leadership competencies. However, several such assignments in a career were de rigueur if one wanted to ascend to IBM senior management.

I broke away from the meeting late that morning and went to Armonk headquarters to have lunch with Jack Kuehler. Jack was president of the company, a member of the board, and John Akers's chief technologist. Kuehler controlled all key technology decisions made in the company. Over lunch he was congenial and easygoing and offered his support. Consistent with my Akers discussion, it was clear that IBM had an obsessive focus on recapturing the ground lost to Microsoft and Intel in the PC world. Jack was almost evangelistic in describing the combined technical strategy behind PowerPC and OS/2—two IBM products that were developed to regain what had been lost to Intel in microprocessors and Microsoft in PC software. The technology plan was sweeping and comprehensive. It sounded exciting, but I had no idea whether it had any chance of succeeding.

After lunch I raced back to the World Trade executive session to hear more about the outlook for our business around the world. In general, it was not good. I then traveled to yet another IBM building to meet with a group of young executives who were in a training class. I returned once again to Armonk to tape a video message for employees, then ended the day with the head of IBM's human resources department, the legendary Walt Burdick.

Burdick had announced his decision to retire before the completion of the CEO search, but I wanted him to stay for at least a short transitional period. IBM's HR department had been well recognized for years for its leadership in many areas, including diversity, recruiting, training, and executive development. Walt Burdick had been in charge of that department for thirteen years, and he was arguably the dean of HR professionals in America.

What perhaps is not as well known is that Burdick was a powerful force behind the throne, one of IBM's highest-paid executives for many years and a major player in creating and enforcing the dominant elements of IBM's culture. His primary interests were structure and process. In fact, after his departure, someone gave me one of the most remarkable documents I have ever seen. Roughly sixty pages long, it is entitled "On Being the Administrative Assistant to W. E. Burdick, Vice President, Personnel, Plans, and Programs." It was written on March 17, 1975, and illustrated some of the suffocating extremes one could find all too easily in the IBM culture. The instructions for an AA in Burdick's office included:

- White shirt and suit jacket at all times.
- Keep a supply of dimes with you. They are helpful when WEB (Burdick) has to make a call when away from the building.
- Surprise birthday parties for WEB staff should be scheduled under the heading "Miscellaneous" for fifteen minutes. Birthday cakes, forks, napkins, and cake knife are handled by WEB's secretary. AA takes seat closest to the door to answer phones.
- WEB has three clocks: one on desk; one on table; one on windowsill outside your office. All three should be reset daily. Call 9-637-8537 for the correct time.
- WEB enjoys Carefree Spearmint sugarless gum. When empty box appears in out-basket, reserve box should be put in his desk and new reserve box purchased.

Burdick and I spent nearly all of our time that day discussing two critical searches that were under way before I had joined IBM: the search for Burdick's replacement, and the search for a Chief Financial Officer (CFO). The prior CFO, Frank Metz, had retired under pressure in January following the same board meeting that had created the CEO search committee. Nothing was more important to me on that first day than filling these jobs. Parachuting into a $65 billion company that was hemorrhaging cash and trying to turn it around is a daunting enough task. Trying to do it without a good CFO and HR director is impossible.

By 6:30 P.M., I finally had the first quiet time of the day. I sat with my longtime assistant, Isabelle Cummins, whom I had talked into coming to IBM despite her desire to retire. Isabelle is an extraordinary person of enormous talents and one of the many heroes of this book. Had she grown up in a later era, she would have been a senior female executive in corporate America, and one of the best. However, that was not the case, and instead she had been my teammate for fifteen years before I came to IBM. I talked her out of retiring because I knew it would have been impossible for me to make it through the early IBM crises—the toughest ones—if she had not been there. At the end of that first day, we shared our experiences and both of us felt totally overwhelmed. (Isabelle, who had always worked with me one-on-one, discovered that nine people, including several AAs and one person responsible for creating and maintaining organization charts, reported to her.)

Early Priorities

The next two weeks were filled with meetings with my direct reports, interviews with candidates for the CFO and HR jobs, and visits to key IBM sites. One of the most important meetings occurred on my second day. I had asked my brother Dick to come by and talk to me about the company. Dick had been a fast-rising star at IBM for many years, having joined the company right out of college. He had served in Europe and, at one point, had headed up the powerful Asia-Pacific region. My guess is that he had been on track to become one of the top executives—a member of the elite and revered Management Committee—but he was tragically cut down by undetected Lyme disease at the height of his career. He had gone on medical leave about six months before John Akers had left, but several executives had asked him to come back and do some consulting for the company. His most important task was working with Nick Donofrio, then head of the Large-Scale Systems Division, to figure out what to do with the mainframe.

Dick (or Rich, as the family has always called him) and I were close as children, he being the oldest and I always the follower in his footsteps—not uncommon, I guess, for two relatively successful siblings. We went our separate ways in adult life, but we always enjoyed each other's company at family gatherings. I never felt any sense of rivalry as each of us climbed the corporate ladder.

Nevertheless, it had to be a poignant moment as he came into the CEO's office at IBM and saw me sitting where, quite realistically, he might have sat had health problems not derailed his career. He came extremely well prepared. In fact, his was the most insightful review anyone had given me during those early days. In particular, he argued against the premise that the mainframe was dead and against a seemingly hysterical preoccupation in the company to allocate all its resources to winning the PC war. I quote directly from the papers he gave me: "We have allowed the info industry to endorse the paradigm that the mainframe is expensive, complex, not responsive, and workstation solutions are cheap, simple to operate, and responsive to business needs. While there is no truth to this paradigm, we have allowed competitors, opinion leaders, and our customers to exaggerate the differences. The result is a dramatic falloff in S/390 (mainframe) sales, increased credibility for Amdahl and Hitachi alternatives, loss of credibility for CIOs (Chief Information Officers) at major corporations, and loss of confidence that IBM had the customers' best interests in mind in its sales organization.

"We should cut the price of hardware ASAP, simplify software pricing, focus development on simplification, implement a hard-hitting communication program to reposition the mainframe and workstations, and underscore that the mainframe is an important part of the CIO's information portfolio."

As I think back on the three or four things that really made a difference in the turnaround of IBM, one of them was repositioning the mainframe. And nobody pointed it out sooner or more clearly than my brother Dick.

He also gave me a few tips that he labeled "brotherly advice":

- Get an office and home PC. Use PROFS (the internal messaging system); your predecessor didn't and it showed.
- Publicly crucify shortsighted proposals, turf battles, and backstabbing. This may seem obvious, but these are an art form in IBM.
- Expect everything you say and do to be analyzed and interpreted inside and outside the company.
- Find a private cadre of advisors who have no axes to grind.
- Call your mom.

Over the next few months I would have liked more advice from Dick, but there was a very watchful group of people at IBM waiting to see if I was setting him up as my own force behind the throne. I didn't want to do that to him or to me. We talked several times, but briefly, and not with the impact his first meeting had on me and the company.

On April 13 I interviewed Jerry York at IBM's office in New York City. Jerry was then Chief Financial Officer at Chrysler Corporation and was one of two candidates I was seeing that week for the CFO job. It was a truly memorable interview. Jerry arrived in a starched white shirt and a blue suit, everything crisp and perfect—West Point style. He was not coy and did not pull any punches. He basically said he wanted the job and then proceeded to outline a series of things that he thought needed to be done as soon as possible. I was impressed by his frankness, his lack of guile, and his candor, as well as his analytical capabilities. It was clear to me that he was tough—very tough—and just what I needed to get at the cost side of IBM. I spoke to another candidate later that week, but I decided Jerry was the right person. He joined us on May 10.

I also saw Gerry Czarnecki, a candidate for the HR job. Gerry was an operations executive at a bank but had been an HR professional years before. Over the next couple of weeks we met several times, both by phone and in person. Although I liked Gerry's energy and his directness, I wasn't certain he was prepared to go back into the HR function. He said, "Probably not anywhere else, but to be part of the turn-around of IBM, I'm prepared to do it."

That turned out to be one of the few hirings that didn't work out as planned during my early years at IBM. It soon became clear to me that it was proving very hard for Gerry to go back and lead the professional HR community. Within four months he appeared to be acting and sounding more like a vice CEO. It wasn't that Gerry's ideas were wrong—in fact, he was a major proponent of substantial cultural change. However, the organization wasn't going to accept from Gerry what it would accept from me. He burned his bridges with his colleagues very soon and he left IBM within a year of his hiring.

Of course, my top priority during those first few weeks was meeting privately with each of the senior executives. A few of them had prepared the ten-page briefing I had requested; most of them offered a more ad hoc analysis of their businesses. In all the meetings over those several weeks, I was sizing up my team, trying to understand the problems they faced and how they were dealing with them, how clearly they thought, how well they executed, and what their leadership potential really was.

The person I relied on most during those early days was Paul Rizzo. As I said earlier, he had been called back from retirement by the board to help John Akers. Paul had been a senior executive at IBM for twenty-two years. After retiring, he became dean of the Business School at the University of North Carolina and was building a new house in that state. The last thing he needed was to come back to IBM, but he did because he loved the company and he didn't want to see it die.

When I arrived, Paul was responsible for the program of federalism—breaking up the company into individual, autonomous units. Not that Paul created the strategy, but in the absence of a CFO, he was basically overseeing the finance function for the board. He was also in charge of watching all the investment bankers who were scrambling over most parts of the company, dollar signs in their eyes as they planted their flags into each business unit. It reminded me of a gold rush. Each one saw an initial public offering of stock (IPO) for the unit or units that he or she advised. We were spending tens of millions of dollars on accountants to create the bookkeeping required for IPOs because IBM's financial system did not support stand-alone units. Paul was also deeply involved in the financing activities that were under way to raise additional capital.

For me, asking Paul to stay on was an easy decision, and I'm grateful he did. Over the next year he was a tower of strength, a wise mentor, and an insightful partner in evaluating strategy and people— another important hero of the IBM turnaround.

A special moment occurred during those first weeks of April. I walked out of my home one morning at my usual early hour. However, when I opened my car door, I suddenly realized there was someone sitting in the back seat. It was Thomas J. Watson, Jr., former IBM CEO and the son of IBM's founder. Tom literally lived across the street and had walked up my driveway to surprise me and ride to work with me. He was 79 years old, and he had retired as CEO of IBM in 1971.

He was animated and, perhaps better stated, agitated. He said he was angry about what had happened to "my company." He said I needed to shake it up "from top to bottom" and to take whatever steps were necessary to get it back on track.

He offered support, urged me to move quickly, reflected on his own career, and, in particular, the need he had seen over and over again to take bold action. At the end of our ride together, I had the feeling he wished he could take on the assignment himself!

On April 15 I made my first official visit to a nonheadquarters site. I had chosen it carefully: the company's research laboratory in Yorktown Heights, New York. If there was a soul of IBM, this lab was it.

Appropriately named the T. J. Watson Research Center, it contained the intellectual fervor that had led IBM over decades to invent most of the important developments that had created the computer industry.

It was my first "public" appearance inside IBM, and it was important because I knew this was my greatest immediate vulnerability. Would the researchers reject me as an unacceptable leader? Some in the company were calling me the "Cookie Monster" because of my previous job at Nabisco.

I spoke from a stage in an auditorium. The house was full, and my remarks were broadcast to an overflow of employees in the cafeteria. Other IBM research facilities around the world picked up the broadcast as well.

The stereotype of researchers says they are so focused on big ideas that they are disconnected from the real world. Well, not these researchers! I saw the pain of IBM's problems on their faces. I don't know if they were curious or apprehensive, but they certainly came to listen.

I gave what soon became my stump speech on focus, speed, customers, teamwork, and getting all the pain behind us. I talked about how proud I was to be at IBM. I underscored the importance of research to IBM's future, but I said we probably needed to figure out ways to get our customers and our researchers closer together so that more of IBM's great foundry of innovation would be aimed at helping people solve real, and pressing, problems.

There was applause, but I wasn't sure what they were thinking.

The Shareholders' Meeting

Perhaps the most traumatic event of my first month at IBM was the annual shareholders' meeting. It had been scheduled, I'm sure, several years in advance for April 26 in Tampa, Florida. Needless to say, it was a daunting challenge to chair my first shareholders' meeting when the company had such major and visible problems. I had been there for only three weeks, could barely identify the products, let alone explain what they did, or, God forbid, describe the technologies inside them. Moreover, it was clear IBM's shareholders were angry and out for blood—perhaps deservedly so. IBM stock had dropped from a high of $43 a share in 1987 to $12 a share the day of the shareholders' meeting. That was less than half its price at the previous year's meeting.

There were 2,300 shareholders waiting impatiently for the show to start when I walked out onto the stage at 10 A.M. that day—in the biggest convention hall I had ever seen. You couldn't help but notice a sea of white hair—obviously, a lot of retirees in Florida owned IBM stock. I made a brief speech in which I asked for some patience, but I made it clear that I was going to move quickly, make all changes necessary, and return the company's focus to the customer.

I got polite applause, and then the fireworks started. Shareholder after shareholder stood up and blasted the company, and frequently the Board of Directors, all of whom were sitting in front of me in the first row of the auditorium. It was a massacre. The directors took direct hit after direct hit. The shareholders were reasonably kind to me in terms of not holding me accountable for the problems, but they also showed little patience for anything other than a fast recovery. It was a long, exhausting meeting—for everyone, I think.

I remember flying back to New York alone that evening on an IBM corporate airplane. My thoughts turned to the Board of Directors. It was clear from the annual meeting that board changes would be necessary—and sooner rather than later. I turned to the flight attendant and said, "This has been a really tough day. I think I'd like to have a drink."

She said, "You don't mean an alcoholic drink, do you?"

"I certainly do!" I replied. "What kind of vodka do you have?"

"We have no alcohol on IBM airplanes. It is prohibited to serve alcohol."

I said, "Can you think of anyone who could change that rule?"

"Well, perhaps you could, sir."

"It's changed, effective immediately."

[4]

Out to the Field

It was crucial that I get out into the field. I didn't want my understanding of the company to be based on the impressions of headquarters employees. Moreover, the local IBM princes and barons were eager to view the new leader. So the day after the annual meeting, I flew to France to meet with the mightiest of all nobles—IBM Europe, Middle East, and Africa (we call it "EMEA"). I visited France, Italy, Germany, and the United Kingdom, all in one week. It was dawn-to-midnight business reviews with senior executives, employee "town hall" meetings, and customer visits.

IBM EMEA was a giant organization operating in 44 countries with more than 90,000 employees. Revenue had peaked at $27 billion in 1990 and had declined since. Gross profit margin on hardware had dropped from 56 percent in 1990 to 38 percent in 1992. Very important was the fact that in the face of this huge decline in gross profits, total expenses had dropped only $700 million. Pretax profit margin had declined from 18 percent in 1990 to 6 percent in 1992.

Wherever I went, the business message was the same: rapidly declining mainframe sales, much higher prices than those of our competitors, a lack of participation in the rapidly growing client/server (PC-centric) segment, and an alarming decline in the company's image. One of the most disturbing statements in my advance reading material was: "We estimate our net cash change at negative $800 million in 1993. We expect to be self-funding but will not be able for some time to pay dividends to the corporation."

While I learned a lot on this trip—the meetings with customers were particularly useful—perhaps the most important messages were internal. It was clear that at all levels of the organization there was fear, uncertainty, and an extraordinary preoccupation with internal processes as the cause of our problems and, therefore, a belief that tinkering with the processes would provide the solutions we needed. There were long discussions of transfer pricing between units, alternative divisions of authority, and other intramural matters. When EMEA executives summarized their action program for the company, number one was: "Use country as prime point of optimization."

I returned home with a healthy appreciation of what I had been warned to expect: powerful geographic fiefdoms with duplicate infrastructure in each country. (Of the 90,000 EMEA employees, 23,000 were in support functions!)

I also came away with an understanding that these were enormously talented people, a team as deeply committed and competent as I had ever seen in any organization. I reached this conclusion repeatedly over the next few months. On the flight home I asked myself: "How could such truly talented people allow themselves to get into such a morass?"

The Click Heard Round the World

As Paul Rizzo had said in our secret meeting in Washington, D.C., IBM's sustainability, at least in the short term, depended heavily on the mainframe. More than 90 percent of the company's profits came from these large "servers" and the software that ran on them. It didn't take a Harvard MBA or a McKinsey consultant to understand that the fate of the mainframe was the fate of IBM, and, at the time, both were sinking like stones.

One of the first meetings I asked for was a briefing on the state of this business. I remember at least two things about that first meeting with Nick Donofrio, who was then running the System/390 business. One is that I drove to his office in Somers, New York, about fifteen miles north of Armonk, and experienced a repeat of my first day on the job. Once again, I found myself lacking a badge to open the doors at this complex, which housed the staffs of all of IBM's major product groups, and nobody there knew who I was. I finally persuaded some kind soul to let me in, found Nick, and we got started. Sort of.

At that time, the standard format of any important IBM meeting was a presentation using overhead projectors and graphics on transparencies that IBMers called—and no one remembers why—"foils." Nick was on his second foil when I stepped to the table and, as politely as I could in front of his team, switched off the projector. After a long moment of awkward silence, I simply said, "Let's just talk about your business."

I mention this episode because it had an unintended, but terribly powerful ripple effect. By that afternoon an e-mail about my hitting the Off button on the overhead projector was crisscrossing the world. Talk about consternation! It was as if the President of the United States had banned the use of English at White House meetings.

By the way, in the telling of that story, I'm in no way suggesting that Nick didn't know his business. In many ways he was the godfather of the technology that would end up saving the IBM mainframe, and his strong technical underpinnings, combined with his uncanny ability to translate technical complexities into common language, were a great source of reassurance to me in the days ahead. We had a great meeting, and there is a straight line between what I heard that day and one early major decision at IBM.

The Mainframe Decision

In a subsequent meeting in the conference room near my office in Armonk, the mainframe team documented a rapid decline in sales and, more important, a precipitous drop in market share in the last fifteen months. I asked why we were losing so much share, and the answer was, "Hitachi, Fujitsu, and Amdahl are pricing 30 to 40 percent below our price."

I asked the obvious: "Why don't we lower our prices so they don't keep beating us like a drum?"

The answer: "We would lose substantial revenues and profits at a time when we need profits badly."

I had hoped to follow the advice of all the management gurus and try to avoid making major decisions in the first ninety days, but that only happens in guru world. The company was hemorrhaging, and at the heart of it was the System/390 mainframe. But almost immediately after joining the company, I had to do something.

It became clear to me at that point that the company, either consciously or unconsciously, was milking the S/390 and that the business was on a path to die. I told the team that, effective immediately, the milking strategy was over and instructed them to get back to me with an aggressive price reduction plan that we could announce two weeks later at a major customer conference.

The financial people gulped hard. There was no doubt that a new CEO could take the alternative strategy: Keep S/390 prices high for a number of years, since it wasn't easy for customers to shift to competitive products in the near future. The revenue—hundreds of millions of dollars—would have been a powerful short-term underpinning of a restructuring of the company. But it would also have been painful for customers and contrary to what they were pleading with us to do, which was to fix the problem rather than walk away from it. Over the longer term, we would have destroyed the company's greatest asset—and perhaps the company itself. So we made a bet on a dramatic price reduction on the product that produced virtually all of IBM's profit.

We made another important decision that day—or, better said, I reaffirmed an important decision that had been made a number of months before I'd arrived. The technical team in the 390 division had staked out a bold move to a totally different technical architecture for the System/390: to move from what was known as a bipolar to CMOS (pronounced "C-moss") technology. If this enormously complex project could be pulled off, it would permit substantial price reductions in the S/390 without commensurate loss in gross profit, thus improving dramatically the competitiveness of the S/390 versus alternative products. If the project failed, the 390 was dead.

But it didn't fail! And the technical wizards from labs in Europe and the United States who pulled it off deserve a place among the heroes of the new IBM. I have always been thankful (and lucky) that some insightful people had made that decision before I'd arrived. My job was simply to reaffirm it and to protect the billion dollars we would spend on it over the next four years.

I am convinced that had we not made the decision to go with CMOS, we'd have been out of the mainframe business by 1997. In fact, that point has been proven more or less by what happened to our principal competitor at the time, Hitachi. It continued development of bigger and bigger bipolar systems, but that technology eventually ran out of gas, and Hitachi is no longer in this business.

The CMOS performance curve was staggering on paper, and it didn't disappoint us. We're building bigger, more powerful systems today than anyone ever dreamed about with bipolar technology. So if you want to think about the return on the $1 billion investment we made back in the early 1990s, I think one fair measure is high-end server revenue from 1997 forward—$19 billion through the end of 2001.

The First Strategy Conference

On Sunday, May 16, I convened a two-day internal meeting on corporate strategy at a conference center in Chantilly, Virginia. There were twenty-six senior IBM executives present. Dress was casual, but the presentations were both formal and formidable.

I was totally exhausted at the end. It was truly like drinking from a fire hose. The technical jargon, the abbreviations, and the arcane terminology were by themselves enough to wear anyone down. But what was really draining was the recognition that while the people in the room were extremely bright, very committed, and, at times, quite convinced of what needed to be done, there was little true strategic underpinning for the strategies discussed. Not once was the question of customer segmentation raised. Rarely did we compare our offerings to those of our competitors. There was no integration across the various topics that allowed the group to pull together a total IBM view. I was truly confused, and that may have been the real low point of my first year at IBM. I walked out of that room with an awful feeling in the pit of my stomach that Murphy and Burke had been wrong— IBM needed a technological wizard to figure out all this stuff!

I didn't have much time to feel sorry for myself because that evening we began what may have been the most important meeting of my entire IBM career: the IBM Customer Forum.

The Customer Meeting at Chantilly

This meeting had been scheduled well before my arrival at the company. Nearly 175 chief information officers of the largest United States companies were coming to hear what was new at IBM. They represented many of the most important customers IBM had—and they could make or break us.

On Tuesday night, I met with several CIOs at dinner, and they shared the same perspective I had heard in Europe. They were angry at IBM—perturbed that we had let the myth that "the mainframe was dead" grow and prosper. The PC bigots had convinced the media that the world's great IT infrastructure—the back offices that ran banks, airlines, utilities, and the like—could somehow be moved to desktop computers. These CIOs knew this line of thinking wasn't true, and they were angry at IBM for not defending their position. They were upset about some other things, too, like mainframe pricing for both hardware and software. They were irritated by the bureaucracy at IBM and by how difficult it was to get integration—integration of a solution or integration across geographies.

Early the next morning, I threw out my prepared speech and decided to speak extemporaneously. I stood before my most important customers and started talking from the heart. I began by telling my audience that a customer was now running IBM; that I had been a customer of the information technology industry for far longer than I would ever be an IBM employee; that while I was not a technologist, I was a true believer that information technology would transform every institution in the world. Thus, I had a strategic view about information technology, and I would bring that to IBM and its customers.

I addressed the issue of the mainframe head-on. I said I agreed with the CIOs that we had failed in our responsibility to define its role in a PC world, that our prices were high, and that there was no question that we were bureaucratic. I shared with them some of my bad experiences with IBM as communicated to me by my CIOs when I was at American Express and RJR Nabisco.

I laid out my expectations:

- We would redefine IBM and its priorities starting with the customer.
- We would give our laboratories free rein and deliver open, distributed, user-based solutions.

- We would recommit to quality, be easier to work with, and reestablish a leadership position (but not the old dominance) in the industry.

- Everything at IBM would begin with listening to our customers and delivering the performance they expected.

Finally, I made the big mainframe pricing announcement. Our team had been working hard over the past two weeks and literally was still putting the proposal together the night before this big meeting. I didn't delve into the details—that was done later in the meeting—but I made it very clear that mainframe prices, both hardware and software, were coming down, and coming down quickly. The price of a unit of mainframe processing moved from $63,000 that month to less than $2,500 seven years later, an incredible 96-percent decline. Mainframe software price/performance improved, on average, 20 percent a year for each of the next six years.

This program, probably more than any other, saved IBM. Over the short term it raised the risk of insolvency as it drained billions of dollars of potential revenue and profits from the company. Had the strategy not worked, I would have been the CEO who had presided over the demise of the company—Louis the Last. However, the plan did work. IBM mainframe capacity shipped to customers had declined 15 percent in 1993. By 1994, it had grown 41 percent, in 1995 it had grown 60 percent, followed by 47 percent in 1996, 29 percent in 1997, 63 percent in 1998, 6 percent in 1999, 25 percent in 2000, and 34 percent in 2001. This represented a staggering turnaround. While pricing was not the only reason IBM survived, it would not have happened had we not made this risky move.

Operation Bear Hug

In late April we had a meeting of the Corporate Management Board. This was the group of fifty top executives with whom I had met in March, the day I was announced as the new CEO.

I shared with them my observations after three weeks on the job. I started by saying that I saw a lot of positive things going on, particularly in research, product development, and in the can-do attitude of a number of people.

However, there were troublesome areas, including:

- Loss of customer trust, supported by some disturbing customer ratings on quality.
- The mindless rush for decentralization, with managers leaping forward saying "make me a subsidiary."
- Cross-unit issues not being resolved quickly.
- Major tension in the organization over who controlled marketing and sales processes.
- A confusing and contentious performance measurement system, causing serious problems when closing sales with customers.
- A bewildering array of alliances that didn't make any sense to me.

I announced Operation Bear Hug. Each of the fifty members of the senior management team was to visit a minimum of five of our biggest customers during the next three months. The executives were to listen, to show the customer that we cared, and to implement holding action as appropriate. Each of their direct reports (a total of more than 200 executives) was to do the same. For each Bear Hug visit, I asked that a one- to two-page report be sent to me and anyone else who could solve that customer's problems. I wanted these meetings to be a major step in reducing the customer perception that dealing with us was difficult. I also made it clear that there was no reason to stop at five customers. This was clearly an exam in which extra credit would be awarded.

Bear Hug became a first step in IBM's cultural change. It was an important way for me to emphasize that we were going to build a company from the outside in and that the customer was going to drive everything we did in the company. It created quite a stir, and when people realized that I really did read every one of the reports, there was quick improvement in action and responsiveness.

The Management Committee Dies

That same day in late April, there was a meeting of the Management Committee (its inside-IBM name was "the MC"). It is important to understand that a seat on the MC was the ultimate position of power that every IBM executive aspired to as the apex of his or her career. When I'd joined the company there were six members, including Akers and Kuehler. The MC met once or twice a week, usually in formal, all-day meetings with lots of presentations. Every major decision in the company was presented to this committee.

Some members of the MC had only recently been appointed. To their utter—and probably crushing—dismay, I told them that afternoon, at my first meeting, that it was unlikely this structure would continue. I wanted to be more deeply involved personally in the decision making of the company, and I was uncomfortable with committees making decisions. While it wasn't officially disbanded until months later, the Management Committee, a dominant element of IBM's management system for decades, died in April 1993.

In some ways, the rise and fall of the Management Committee symbolized the whole process of rigor mortis that had set in at IBM. It seemed to me an odd way to manage a company—apparently centralized control, but in a way that ultimately diffused responsibility and leadership. The MC was part of IBM's famed contention system, in which the recommendations of powerful line units were contested by an equally powerful corporate staff. As I think about the complexity of the technology industry and the risks associated with important business and product decisions, this approach may very well have been a brilliant innovation when it was created. The problem was that over time, IBM people learned how to exploit the system to promote their own agendas. So by the early 1990s a system of true contention was apparently replaced by a system of prearranged consensus. Rather than have proposals debated, the corporate staff, without executives, worked out a consensus across the company at the lowest possible level. Consequently, what the Management Committee most often got to see was a single proposal that encompassed numerous compromises. Too often the MC's mission was a formality—a rubber-stamp approval.

I haven't spent much time unearthing and analyzing IBM's history, but I have been told that the administrative assistant network emerged as the facilitator of this process of compromise. Much like the eunuchs of the ancient Chinese court, they wielded power beyond their visible responsibilities.

Meetings with Industry Experts

In the course of everything else during the first weeks of my being on the job, I scheduled a number of one-on-one meetings with various leaders in the computer and telecommunications industry. They included John Malone of TCI, Bill Gates of Microsoft, Andy Grove of Intel, Chuck Exley of NCR, and Jim Manzi of Lotus. These meetings were very helpful to me, more for their insights into the industry than for anything said about IBM. And, as you might expect, many of my visitors arrived with thinly disguised agendas.

The meeting with Andy Grove was perhaps the most focused. In his wonderfully direct style, Andy delivered the message that IBM had no future in the microprocessor business, that we should stop competing with Intel with our PowerPC chip, and that, unless this happened, relationships between the two companies were going to be difficult. I thanked Andy, but, having no real understanding at that point of what we should do, I tucked the message away.

The meeting with Bill Gates was not significant from the point of view of content. Basically, he delivered the message that I should stick to mainframes and get out of the PC business. More memorable are the incidentals.

We met at 8 A.M. on May 26 at the IBM building on Madison Avenue in New York City. Coincidentally, I was to meet later that same day with Jim Manzi, head of Lotus. The IBM security person in the lobby got confused and called Gates "Mr. Manzi" and gave him Manzi's security pass. By the time Bill arrived on the 40th floor, he wasn't happy. Nevertheless, we had a useful discussion.

What followed the meeting was more noteworthy. He and I, as well as our staffs, had agreed there would be no publicity in advance of or after the meeting. However, the press had the story two hours after he left the IBM building, and by evening everyone knew about the confusion over his security badge. He apparently didn't deduce that I had a meeting with Manzi that same day. To some, the mix-up seemed to be further evidence of IBM's—and perhaps Lou Gerstner's—ineptitude.

The Financials: Sinking Fast

We announced first-quarter operating results at the end of April, and they were dismal. Revenue had declined 7 percent. The gross profit margin had fallen more than 10 points—to 39.5 percent from 50 percent. The company's loss before taxes was $400 million. In the first quarter of the previous year, IBM had had a pretax profit of close to $1 billion.

At the end of May I saw April's results, and they were sobering. Profit had declined another $400 million, for a total decline of $800 million for the first four months. Mainframe sales had dropped 43 percent during the same four months. Other large IBM businesses—software, maintenance, and financing—were all dependent, for the most part, on mainframe sales and, thus, were declining as well. The only part of the company that was growing was services, but it was a relatively small segment and not very profitable. Head count had declined slightly, from 302,000 at the beginning of the year to 298,000 at the end of April. Several business units, including application-specific software and our semiconductor businesses, were struggling.

Almost as frustrating as the bad results was the fact that, while the corporation could add up its numbers quite well in total, the internal budgeting and financial management systems were full of holes. There was not one budget but two or three, because each element of the IBM organization matrix (e.g., the geographic units versus the product divisions) insisted on its own budget. As a result, there really wasn't a single, consolidated budget. Allocations were constantly debated and changed, and accountability was extremely difficult to determine.

Given the fact that the mainframe was still in free fall and so much of IBM's business at that point depended on the mainframe, the outlook was extremely precarious. We were shoring up the balance sheet as best we could with financing, but something had to be done to stabilize the operations.

The Media Early On

There was a very short honeymoon with the media—understandably, given the nature of the story, but also because it's impossible to transform a badly ailing company under the glare of daily press briefings and publicity. There's too much work to do inside without having to contend with a daily progress report in the papers focusing everyone on results that take months and years, and not hours and days, to achieve. A reporter from the Associated Press wanted to follow me around all day my first day. *USA Today* said it was working on graphics for a daily progress chart. We said, "No, thank you. We're going dark for a bit while we assess the task at hand." That was not a popular way to answer reporters who were used to writing daily stories about the problems at IBM.

I brought with me to IBM from the first day my communications executive, David Kalis. David had been with me for many years, going back to American Express in the 1980s. He was, in my opinion, the best public relations executive in America. He was also the first true PR professional in IBM's history to hold the top communications job. For decades the position had been a rotation slot for sales executives being groomed for other top jobs.

He inherited a shambles at IBM. There were some talented people, but the communications department was staffed for the most part with well-meaning but untrained employees. However, even if they had all been professionals, it would have been impossible for them to perform, given the foxhole mentality that permeated the company in

1993. Typically, IBM executives believed that the only real problem the company had was the daily beating it was getting in the press. They felt that if we had more positive stories in the media, IBM would return to profitability and everything would be normal again.

Although my clear priority was meeting and talking with IBM customers and employees, I had to make some time for the media as well. Pressure was almost overwhelming. During the first months, on any given day there were more than sixty-five standing requests for Gerstner interviews from the major media. If requests from local newspapers and computer industry press were factored in, the demand was in the hundreds. If one included international demand . . . and so on.

I did what I could on a tight schedule. I conducted individual interviews with *The New York Times, The Wall Street Journal, Business-Week, Fortune, USA Today,* and the *Financial Times.* But, the media let us know, loudly and frequently, that that just wasn't enough.

More than anything else, I wanted time, but I knew I didn't have a lot of it. Pressure was building—in the media, on the Street, and with shareholders. A lot had been done, but I knew I would have to go public, and do it soon, with my plans to fix IBM.

[6]

Stop the Bleeding
(and Hold the Vision)

B y July 1993, the pressure to act—and to act in a comprehensive manner—was acute. The financials were ominous. Employees wanted their new leader to do something, anything, to give them a sense of direction. The media, typically, were losing patience (not that I've ever felt the media had any particularly noteworthy insights into what was going on in IBM, but given the fragility of the company at that time, their stories, right or wrong, could have had a devastating impact on customer attitudes).

On July 14, *USA Today* celebrated my one hundredth day on the job in a long cover story. This was the story's lead:

> IBM stockholders and customers might have hoped for miracles in Louis Gerstner's first 100 days as IBM's CEO. But that honeymoon period ended Friday—with no major organizational overhauls or strategic moves.

"Clearly he is not a miracle worker," computer analyst Ulric Weil says.

IBM stock, down 6% since Gerstner took over, "has done nothing because he's done nothing," says computer analyst David Wu at S. G. Warburg.

Although I thought I had already done a lot, it was clearly time to make some major decisions and go public with them. And after all the customer and employee and industry meetings, as well as weekend and air travel reflection, I was indeed ready to make four critical decisions:

- Keep the company together.
- Change our fundamental economic model.
- Reengineer how we did business.
- Sell underproductive assets in order to raise cash.

And, I decided I would not disclose any of the other strategic initiatives that were forming in my head during the summer of 1993.

Keep the Company Together

I can't tell you exactly when I decided to keep IBM together, nor do I remember a formal announcement. I had always talked about our size and breadth as a distinct competitive advantage. However, I do know that it wasn't a particularly difficult decision for me. Here's why.

When the computer industry first appeared on the world's stage, its model was to deliver to customers a total, integrated package. When a company bought a computer, it came with all the basic technologies, like microprocessors and storage, incorporated into a system; all the software loaded onto the hardware; all the services to install and maintain the system were bundled into the pricing. The customer basically purchased a total system and had it installed for a single price. This was the model created by IBM, and over time, only a handful of computer competitors, all fully integrated, emerged in the United States (often described as the BUNCH—Burroughs, Univac, NCR, Control Data, and Honeywell). The same model emerged in Japan and, to a lesser extent, in Europe.

In the mid-1980s a new model started to appear. It argued that vertical integration was no longer the way to go. The new breed of successful information technology companies would provide a narrow, horizontal slice of the total package. So companies that sold only databases began to emerge, as well as companies that sold only operating systems, that sold only storage devices, and so on. Suddenly the industry went from a handful of competitors to thousands and then tens of thousands, many of which sold a single, tiny piece of a computer solution.

It was in this new environment that IBM faltered, and so it was logical for many of the visionaries and pundits, both inside and outside the company, to argue that the solution lay in splitting IBM into individual segments. This conclusion, however, appeared to me to be a knee-jerk reaction to what new competitors were doing without understanding what created fragmentation in the industry.

Two things really drove the customer to support this new, fragmented supplier environment:

- Customers wanted to break IBM's grip on the economics of the industry—to rip apart IBM's pricing umbrella, which allowed it to bundle prices and achieve significantly high margins.
- The customer was increasingly interested in delivering computing power to individual employees (the term was "distributed computing," in contrast to the mainframe's "centralized computing").

IBM was slow, very slow, in delivering distributed computing, and many small companies moved in to fill the gap. These companies were in no position to deliver an entire integrated solution, so they offered add-ons to the basic IBM system and built around IBM's central processing hub. This is clearly what Microsoft and Intel did when IBM reluctantly moved into the PC business.

So it really wasn't that the customer desired a whole bunch of fragmented suppliers. The broad objective was to bring more competition into the marketplace and to seek suppliers for a new model of computing.

And it worked. By the early 1990s there were tens of thousands of companies in the computer industry, many of which lived for a few months or years, then disappeared. But the impact of all this dynamism was lower prices and more choices (with the notable exception of the PC industry, where Microsoft re-created the IBM choke hold, only this time around it was in the desktop operating environment rather than in the mainframe computer).

While there were good consequences to this fundamental reshaping of the computer industry, there was also one very undesirable outcome. The customer now had to be the integrator of the technology into a usable solution to meet his or her business requirements. Before, there was a general contractor called IBM or Burroughs or Honeywell. Now, in the new industry structure, the burden was on the customer to make everything work together.

This was all complicated by the absence of uniform standards throughout the computer industry. Competitors in the computer industry, unlike any other industry I know, try to create unique standards that they "own." They do not make it easy for other companies in the industry to connect with or understand these standards without extracting a huge price. Thus, there is a cacophony of standards and interconnect requirements that makes the creation of a single solution very difficult. (I'll discuss this in more detail later.)

As a big customer of the information technology industry in the early 1990s, I knew firsthand that integration was becoming a gigantic problem. At American Express, our wallet-sized piece of plastic was going to be moving data all over the world, a fact which brought with it enormous technical challenges. All I wanted was an information technology platform, and a partner, that would allow me to run that business the way I wanted it to run. So when I arrived at IBM in 1993, I believed there was a very important role for some company to be able to integrate all of the pieces and deliver a working solution to the customer.

Why? Because at the end of the day, in every industry there's an integrator. Sure, there are supply chains, and there are enterprises at various points in the chain that offer only one piece of a finished product: steelmakers in the auto industry; component makers in consumer electronics; or providers of a marketing or tax application in financial services. But before the components reach the consumer, somebody has to sit at the end of the line and bring it all together in a way that creates value. In effect, he or she takes responsibility for translating the pieces into value. I believed that if IBM was uniquely positioned to do or to be anything, it was to be that company.

Another myth that was playing out at the time was that the information technology (IT) industry was going to continue to evolve—or devolve—toward totally distributed computing. Everything was going to get more local, more self-contained, smaller and cheaper until the point at which all the information in the world would be running on somebody's wristwatch. A lot of people had bought into the value of information democratization extended to its extreme, and they accepted the industry's promise that all the piece parts would work together, or, in the industry's term, "interoperate."

But even before I had crossed the threshold at IBM, I knew that promise was empty. I'd spent too long a time on the other side. The idea that all this complicated, difficult-to-integrate, proprietary collection of technologies was going to be purchased by customers who would be willing to be their own general contractors made no sense.

Unfortunately, in 1993 IBM was rocketing down a path that would have made it a virtual mirror image of the rest of the industry. The company was being splintered—you could say it was being destroyed.

Now, I must tell you, I am not sure that in 1993 I or anyone else would have started out to create an IBM. But, given IBM's scale and broad-based capabilities, and the trajectories of the information technology industry, it would have been insane to destroy its unique competitive advantage and turn IBM into a group of individual component suppliers—more minnows in an ocean.

In the big April customer meeting at Chantilly and in my other customer meetings, CIOs made it very clear that the last thing in the world they needed was one more disk drive company, one more operating system company, one more PC company. They also made it clear that our ability to execute against an integrator strategy was nearly bankrupt and that much had to be done before IBM could provide a kind of value that we were not providing at the time—but which they believed only IBM had a shot at delivering: genuine problem solving, the ability to apply complex technologies to solve business challenges, and integration.

So keeping IBM together was the first strategic decision, and, I believe, the most important decision I ever made—not just at IBM, but in my entire business career. I didn't know then exactly how we were going to deliver on the potential of that unified enterprise, but I knew that if IBM could serve as the foremost integrator of technologies, we'd be delivering extraordinary value.

As a result, we threw out the investment bankers who were arranging IPOs of all the pieces of the enterprise. We threw out the accountants who were creating official financial statements required in order to sell off the individual components. We threw out the naming consultants who had decided that the printer division should be called "Pennant" and the storage division "AdStar."

We stopped all the internal activities that were creating separate business processes and systems for each of these units, all of which were enormous drains of energy and money. For example, even in the midst of financial chaos, we had managed to hire more than seventy different advertising agencies in the United States alone (more on this later). Human resources people were willy-nilly changing benefits programs so that if an employee left one IBM business unit for another, it was like entering another country, with different language, currency, and customs.

I began telling customers and employees that IBM would remain one unified enterprise. I remember that the response from our executive team was mixed—great joy from those who saw the company as being saved, and bitter disappointment from those who saw a breaking apart as their personal lifeboat to get off the *Titanic*.

Changing Our Economic Model

The second major decision that summer was to restructure IBM's fundamental economics. At the risk of sounding pompously tutorial, a profit-making business is a relatively simple system. You need to generate revenue, which comes from selling things at an acceptable price. You have to achieve a good gross profit on those sales. And you have to manage your expenses, which are the investments you make in selling, research and development, building plants and equipment, maintaining financial controls, developing and running advertising, and so on. If revenue, gross profit, and expenses are all moving in the right relationship, the net effect is growing profits and positive cash flow.

Unfortunately, in IBM's case the relationships were all wrong. Revenue was slowing because the company was so dependent on the mainframe, and mainframe sales were declining. Gross profit margin was sinking like a stone because we had to reduce mainframe prices in order to compete. The only way to stabilize the ship was to ensure that expenses were going down faster than the decline in gross profit.

Expenses were a major problem. After months of hard work, CFO Jerry York and his team determined that IBM's expense-to-revenue ratio—i.e., how much expense is required to produce a dollar of revenue—was wildly out of range with those of our competitors. On average, our competitors were spending 31 cents to produce $1 of revenue, while we were spending 42 cents for the same end. When we multiplied this inefficiency times the total revenue of the company, we discovered that we had a $7 billion expense problem!

Since the repositioning of the mainframe was a long-term challenge and we had to reduce mainframe prices and thus our gross profit, the only way to save the company, at least in the short term, was to slash uncompetitive levels of expenses.

So we made the decision to launch a massive program of expense reduction—$8.9 billion in total. Unfortunately, this necessitated, among other things, reducing our employment by 35,000 people, in addition to the 45,000 people whom John Akers had already laid off in 1992. That meant additional pain for everyone, but this was a matter of survival, not choice.

Reengineer How We Did Business

These early expense cuts were necessary to keep the company alive, but I knew they were far from sufficient to create a vibrant, on-going, successful enterprise. We needed fundamental change in the way we carried out almost every process at IBM. All of our business processes were cumbersome and highly expensive. So in 1993 we began what ultimately became one of the largest, if not *the* largest, reengineering projects ever undertaken by a multinational corporation. It would last a decade and, as it unfolded, change almost every management process inside IBM.

Reengineering is difficult, boring, and painful. One of my senior executives at the time said: "Reengineering is like starting a fire on your head and putting it out with a hammer." But IBM truly needed a top-to-bottom overhaul of its basic business operations.

Jerry York led the effort. By addressing some of the obvious excesses, he had already cut $2.8 billion from our expenses that year alone. Beyond the obvious, however, the overall task was enormous and daunting. We were bloated. We were inefficient. We had piled redundancy on top of redundancy.

We were running inventory systems, accounting systems, fulfillment systems, and distribution systems that were all, to a greater or lesser degree, the mutant offspring of systems built in the early mainframe days and then adapted and patched together to fit the needs of one of twenty-four independent business units. Today IBM has one Chief Information Officer. Back then we had, by actual count, 128 people with CIO in their titles—all of them managing their own local systems architectures and funding home-grown applications.

The result was the business equivalent of the railroad systems of the nineteenth century—different tracks, different gauges, different specifications for the rolling stock. If we had a financial issue that required the cooperation of several business units to resolve, we had no common way of talking about it because we were maintaining 266 different general ledger systems. At one time our HR systems were so rigid that you actually had to be fired by one division to be employed by another.

Rather than approach our reengineering sequentially, we attacked the entire organization at once. At any given time, more than sixty major reengineering projects were under way—and hundreds more among individual units and divisions.

Most of the work centered on eleven areas. The first six we called the core initiatives, meaning those parts of the business that dealt most with the outside world: hardware development, software development (later, these two units were combined into integrated product development), fulfillment, integrated supply chain, customer relationship management, and services.

The rest focused on internal processes, called enabling initiatives: human resources, procurement, finance, real estate, and surprisingly—at least at first glance—information technology.

When I'd arrived at IBM, I wasn't taking too much for granted, but I did expect I'd find the best internal IT systems in the world. This might have been my greatest shock. We were spending $4 billion a year on this line item alone, yet we didn't have the basic information we needed to run our business. The systems were antiquated and couldn't communicate with one another. We had hundreds of data centers and networks scattered around the world; many of them were largely dormant or being used inefficiently.

We saved $2 billion in IT expenses by the end of 1995. We went from 155 data centers to 16, and we consolidated 31 internal communications networks into a single one.

Real estate was an especially big project. The real estate and construction division in the United States had grown so big that it could have been a separate company. In the early 1990s it employed 240 people. We had tens of millions of square feet in fancy city-center office buildings that had been built by IBM in its heyday in the 1970s and 1980s. You couldn't walk into a major city in the United States that did not have a large IBM tower. The same was true overseas. However, by the 1990s, many properties were underused or being rented out. At the same time, we were renting an entire floor in a midtown Manhattan office building mostly for product introductions, for $1 million a year.

We sold 8,000 acres of undeveloped land. We sold first-class real estate that we didn't need, like the tallest building in Atlanta. We hired outside providers and cut full-time staff to 42. Around our corporate headquarters in Westchester County, New York, we consolidated twenty-one locations into five.

From 1994 to 1998, the total savings from these reengineering projects was $9.5 billion. Since the reengineering work began, we've achieved more than $14 billion in overall savings. Hardware development was reduced from four years to an average of sixteen months— and for some products, it's far faster. We improved on-time product delivery rates from 30 percent in 1995 to 95 percent in 2001; reduced inventory carrying costs by $80 million, write-offs by $600 million, delivery costs by $270 million; and avoided materials costs of close to $15 billion.

Sell Unproductive Assets to Raise Cash

The fourth action program that we kicked off that summer represented a scramble to sell unproductive assets and raise cash. Only a handful of people understand how precariously close IBM came to running out of cash in 1993. Whether we would have had to file for bankruptcy, I can't say. There were certainly lots of assets that could be sold to make the company solvent again. The issue was: Could that be done before we turned down that horrible spiral that companies enter when their cash flow shrinks and their creditors are no longer willing to stand behind them?

In July we announced that we would cut our annual dividend to shareholders from $2.16 to $1. That fall, Jerry York and his team went to work to sell any asset that was not essential to the company. We sold much of the corporate airplane fleet. We sold the corporate headquarters in New York City. We had massive investments in expensive training centers where we housed and fed tens of thousands of people a year. In 1993 we had four such private facilities within an hour's drive of the Armonk headquarters, one of which, a former estate of the Guggenheim family on the Gold Coast of Long Island, was used almost exclusively by the IBM human resources organization.

Over the prior decade, IBM had amassed a large and important fine-art collection, most of which was stored in crates out of sight from anyone. Some of it did show up now and again in a public gallery in the IBM tower on 57th Street in Manhattan. We had a curator and a staff who maintained this collection. In 1995 the bulk of it was sold at auction at Sotheby's for $31 million. Unfortunately, the sale was condemned by many people in the art world. For some reason, these people felt that it was fine for IBM to fire employees and send them home, as long as we kept some paintings in a gallery in New York City for people to view occasionally.

The largest sale we made in the first year was IBM's Federal Systems Company, which did major projects primarily for the United States government. We took advantage of the fact that the United States defense industry was consolidating rapidly at that time and there were buyers who would pay top dollar to increase their concentration. The unit had an illustrious history of important technological breakthroughs for various national security and space programs. However, it was also a perpetual low-margin business because we never figured out how to fit it into the overall high-expense model of the commercial side of the business. Loral Corporation bought it in January 1994 for $1.5 billion.

The program of selling off unproductive assets continued for many years. The need to raise cash became less important as we moved into 1995 and 1996. However, as the years went by, we continued to streamline the company for a different reason: focus. (I will return to this subject later.)

Hold the Vision

I've had a lot of experience turning around troubled companies, and one of the first things I learned was that whatever hard or painful things you have to do, do them quickly and make sure everyone knows what you are doing and why. Whether dwelling on a problem, hiding a problem, or dribbling out partial solutions to a problem while you wait for a high tide to raise your boat—dithering and delay almost always compound a negative situation. I believe in getting the problem behind me quickly and moving on.

There were so many constituencies involved with a stake in IBM's future that we decided the only way to communicate with them all about our decisions, including expense reductions and additional layoffs, was through a press conference.

We made the announcement the morning of July 27 at a large meeting room in a midtown Manhattan hotel. It seemed there were two hot topics that year that guaranteed big press attendance—jobs and IBM. So when IBM made an announcement about jobs, it was a full house, with every TV network and major paper in the world present.

This was basically my coming-out event—my first public discussion of what I had learned and planned to do at IBM. I worked hard on what I would say, but given IBM's longtime image of starchy formality, I decided to speak without notes or even a podium. No props. Nothing to lean on. Just me and what I had to say.

I said something at the press conference that turned out to be the most quotable statement I ever made:

"What I'd like to do now is put these announcements in some sort of perspective for you. There's been a lot of speculation as to when I'm going to deliver a vision of IBM, and what I'd like to say to all of you is that the last thing IBM needs right now is a vision."

You could almost hear the reporters blink.

I went on: "What IBM needs right now is a series of very tough-minded, market-driven, highly effective strategies for each of its businesses—strategies that deliver performance in the marketplace and shareholder value. And that's what we're working on.

"Now, the number-one priority is to restore the company to profitability. I mean, if you're going to have a vision for a company, the first frame of that vision better be that you're making money and that the company has got its economics correct.

"And so we are committed to make this company profitable, and that's what today's actions are about.

"The second priority for the company," I said, "is to win the battle in the customers' premises. And we're going to do a lot of things in that regard, and again, they're not visions—they're people making things happen to serve customers."

I continued: "Third, in the marketplace, we are moving to be much more aggressive in the client/server arena. Now, we do more client/server solutions than anybody else in the world, but we have been sort of typecast as the 'mainframe company.' Well, we are going to do even more in client/server.

"Fourth, we are going to continue to be, in fact, the only full-service provider in the industry, but what our customers are telling us is they need IBM to be a full-solutions company. And we're going to do more and more of that and build the skills to get it done.

"And lastly, we're doing a lot of things that I would just call 'customer responsiveness'—just being more attentive to the customer, faster cycle time, faster delivery time, and a higher quality of service."

The reaction to the expense cuts was, for the most part, supportive. "This is the most realistic restructuring program IBM's had," analyst David Wu told *The Wall Street Journal.*

Michael Hammer, coauthor of *Reengineering the Corporation,* told *The New York Times:* "Gerstner decided that sooner is better than perfect—that was anathema to the old IBM. That is the most important kind of change that can come from the top."

As for my vision statement, the doom industry had a grand time nailing my hide to the wall.

The ubiquitous Charles Ferguson, coauthor of *Computer Wars,* told *The New York Times:* "Gerstner can be fabulous at this cost-cutting stuff and still see IBM all but collapse over the next five years or so. The tough part will be deciding what strategies it will pursue and to carve out a profitable niche for IBM in the future."

Barron's was blunt: "George Bush would have called it the vision thing. Others may be calling it the 'lack-of-vision thing,' in a pointed reference to IBM Chairman Louis V. Gerstner's assertion that 'the last thing IBM needs now is a vision.' Instead, he told reporters last week, the company most needs 'market-driven . . . strategies in each of its businesses.' In other words, a super-duper tool kit.

"In truth, however, IBM's newish chief *does* have a solution for the ailing computer colossus, though it is neither poetic nor grand. Rather, it is one of corporate anorexia."

The Economist asked: "But does cost-cutting amount to a strategy for survival?"

The Economist called my intention to keep IBM together "short-sighted." The magazine said: "As PCs become cheaper, more powerful, and easier to link into networks, the number of customers prepared to buy everything from IBM will dwindle. Indeed, IBM's various businesses would be much stronger competitors if they were not hamstrung either by Big Blue's still-vast corporate overheads, or by the need not to tread on other divisions' toes. It may take a few more quarters of leaping losses to convince Mr. Gerstner of the need to break up IBM. Shareholders, their investment at an eighteen-year low, and their dividend halved for the second time this year, might wish that their axeman would turn visionary overnight."

I don't know if I was surprised at the reaction—I guess I was. I was certainly annoyed—and for good reason.

A lot of reporters dropped the words "right now" from my vision statement when they wrote their stories. And so they had me saying that "the last thing IBM needs is a vision." That was inaccurate reporting, and it changed my message in a big way.

I said we didn't need a vision *right now* because I had discovered in my first ninety days on the job that IBM had file drawers full of vision statements. We had never missed predicting correctly a major technological trend in the industry. In fact, we were still inventing most of the technology that created those changes.

However, what was also clear was that IBM was paralyzed, unable to act on any predictions, and there were no easy solutions to its problems. The IBM organization, so full of brilliant, insightful people, would have loved to receive a bold recipe for success—the more sophisticated, the more complicated the recipe, the better everyone would have liked it.

It wasn't going to work that way. The real issue was going out and making things happen every day in the marketplace. Our products weren't bad; our people were good people; our customers had long, successful relationships with us. We just weren't getting the job done. As I said frequently to IBMers those days, "If you don't like the pain, the only answer is to move the pain onto the backs of your competitors. They're the ones who have taken your market share. They're the ones who have taken away your net worth. They're the ones who made it more difficult to send your children and grandchildren to college. The answer is to shift the pain to them and return IBM to a world of success."

Fixing IBM was all about execution. We had to stop looking for people to blame, stop tweaking the internal structure and systems. I wanted no excuses. I wanted no long-term projects that people could wait for that would somehow produce a magic turnaround. I wanted—IBM needed—an enormous sense of urgency.

What the pundits also missed was that we had already made some very fundamental strategic decisions that were the early elements of a vision. I did not talk about them in that July meeting—at least, I didn't talk about them as forthrightly as I might have—because I did not want our competitors to see where we were going. The key strategic decisions that were already made before that eventful day were extraordinarily significant in the turnaround of IBM. They were:

- Keep the company together and not spin off the pieces.
- Reinvest in the mainframe.
- Remain in the core semiconductor technology business.
- Protect the fundamental R&D budget.
- Drive all we did from the customer back and turn IBM into a market-driven rather than an internally focused, process-driven enterprise.

You could argue that a lot of these decisions represented a return to IBM's Watson roots. However, to have announced in July 1993 a strategy built around past experience would have subjected us to gales of laughter that would have blown around the world. If the last thing IBM needed in July 1993 was a vision, the second last thing it needed was for me to stand up and say that IBM had basically everything right and we would stand pat but work harder. That would have had a devastating effect on all our constituents—customers, employees, and shareholders.

So the truly unique challenge of my first few months at IBM was to reject the knee-jerk responses that would have destroyed the company, and to focus on day-to-day execution, stabilizing the company while we sought growth strategies that would build on our unique position in the industry. Those were not to come until a year later.

[7]

Creating the Leadership Team

As 1993 drew to a close, I turned my attention increasingly to the overall IBM team, my top management team, and our Board of Directors.

If you ask me today what single accomplishment I am most proud of in all my years at IBM, I would tell you it is this—that as I retire, my successor is a longtime IBMer, and so are the heads of all our major business units.

I think it would have been absolutely naïve—as well as dangerous—if I had come into a company as complex as IBM with a plan to import a band of outsiders somehow magically to run the place better than the people who were there in the first place. I've entered other companies from the outside, and based on my experience, you might be able to pull that off at a small company in a relatively simple industry and under optimal conditions. It certainly wasn't going to work at IBM. It was too big and too complex a structure. More important, the company was brimming with talented people who had unique expertise. If I didn't give the players on the home team a chance, they'd simply take their talent and knowledge and go some-where else. I just had to find the teammates who were ready to try to do things a different way.

We had many big-stakes business decisions to make, so deciding whom I was going to trust was critically important. There is no easy way to do this. Building a management team is something you have to do business by business, person by person, day by day. I read their re-ports. I watched them interact with customers. I sat with them in meetings and evaluated the clarity of their thinking and whether they had the courage of their convictions or were weathervanes ready to shift direction if I scowled or raised an eyebrow. I needed to know they were comfortable discussing their business problems candidly with me.

When I disbanded the Management Committee during my first month, it was a loud statement that there were going to be major changes in the managerial culture of IBM. However, I still needed a top-level executive committee to work with me to run the company, so in September I created the Corporate Executive Committee, which overnight was widely renamed "the CEC." It had eleven members, in-cluding myself.

With an eye to the old Management Committee, I also announced what the CEC would *not* do: It would not accept delegation of problem solving. It would not sit through presentations or make decisions for the business units. Its focus would be solely on policy issues that cut across multiple units.

It wasn't long before the company's culture decided that the CEC had fully replaced the MC as the ultimate honorific the company could bestow. I have never viewed getting a seat on a committee as some-thing a successful person should truly value. However, sometimes you have to work within the existing system. If all the talented IBMers wanted to work harder in order to get a seat on the CEC, under the cir-cumstances that was okay with me.

At the same time, I created a Worldwide Management Council

(WMC) to encourage communication among our businesses. The WMC had thirty-five members and was to meet four or five times a year in two-day sessions to discuss operating unit results and company-wide initiatives. In my mind, however, its primary purpose was to get the executive team working together as a group with common goals— and not to act as some United Nations of sovereign countries. These meetings represented a chance for our top executives to grab one another and say "I've got a great idea, but I need your help."

Building a New Board

One of the most revolutionary, but least noticed, changes in the early days involved the Board of Directors. When I arrived there were eighteen directors, including four insiders: John Akers, Jack Kuehler, John Opel (IBM's CEO before Akers), and Paul Rizzo. I thought this was an unwieldy size with too many insiders, particularly given the dominance of current and former employees on the powerful Executive Committee.

Clearly the CEO search, the media's public flogging of the company, and the sharp, extended criticism at the annual meeting had traumatized many members of the board. I quietly approached a few of them, especially Jim Burke and Tom Murphy, for a series of discussions on corporate governance.

With my encouragement, the Directors' Committee decided it would announce that the board should be reduced in size to make it more manageable. At the same time, we would add new people to bring in some different perspectives. After the announcement, it didn't take anyone more than a minute to realize that meant a significant amount of retirements would be in order.

I think most of the directors had mixed feelings about sticking around at that point, and some welcomed the opportunity for a graceful exit. Burke and Murphy masterfully orchestrated a proposal that every director offer his or her resignation and that the Directors' Committee would sort out the right structure for the ongoing board.

As a result, five directors left in 1993, then four more in 1994. Murphy and Burke themselves retired, one year earlier than required by IBM's retirement rules. Their move was a sign to the others that it was time to make room for the newcomers. A few were willing to go, but others found the process distasteful and personally difficult. Nevertheless, we got it all done. To the amazement of everyone, there was never so much as a peep in the media.

By the end of 1994 we had a twelve-member board. I was the only insider. Only eight remained from the eighteen who had made up the board just a year before.

Starting in 1993 we began introducing newcomers, beginning with Chuck Knight, the chairman and CEO of Emerson Electric Co. I had known Chuck as a fellow board member at Caterpillar. He was tough and demanding of himself, the CEO, and his fellow board members, and I admired that. He was highly respected as one of the premier CEOs in America, and his selection was the important first step in the rebuilding of the board.

In 1994 we added Chuck Vest, president of MIT and Alex Trotman, chairman and CEO of Ford Motor Company. Cathie Black, president and CEO of the Newspaper Association of America, and Lou Noto, chairman and CEO of Mobil Corporation, joined in 1995. They were followed by Juergen Dormann, chairman of Hoechst AG, in 1996. Minoru Makihara, president of Mitsubishi Corporation and one of the most senior business executives in Japan, joined us in 1997; Ken Chenault, president and chief operating officer (and later chairman and CEO) of American Express in 1998; and Sidney Taurel, chairman and CEO of Eli Lilly and Company in 2001.

This board has been an important contributor to our success. Strong, involved, effective, it has consistently practiced corporate governance in a manner that meets the most rigorous standards. In fact, in 1994 the California Public Employees' Retirement System

(CalPERS), whose board manages one of the world's largest public pension funds, rated the IBM board's governance practices among the very best. Since then there has been noteworthy recognition from other organizations.

Employee Communications

At the same time we were remaking our board and senior management system, it was essential to open up a clear and continuous line of communications with IBM employees. The sine qua non of any successful corporate transformation is public acknowledgment of the existence of a crisis. If employees do not believe a crisis exists, they will not make the sacrifices that are necessary to change. Nobody likes change. Whether you are a senior executive or an entry-level employee, change represents uncertainty and, potentially, pain.

So there must be a crisis, and it is the job of the CEO to define and communicate that crisis, its magnitude, its severity, and its impact. Just as important, the CEO must also be able to communicate how to end the crisis—the new strategy, the new company model, the new culture.

All of this takes enormous commitment from the CEO to communicate, communicate, and communicate some more. No institutional transformation takes place, I believe, without a multi-year commitment by the CEO to put himself or herself constantly in front of employees and speak in plain, simple, compelling language that drives conviction and action throughout the organization.

For me at IBM this meant, in some respects, seizing the microphone from the business unit heads, who often felt strongly about controlling communications with "their people"—to establish their priorities, their voice, their personal brand. In some companies, at some times, such action may be appropriate—but not at the Balkanized IBM of the early 1990s. This was a crisis we *all* faced. We needed to start understanding ourselves as one enterprise, driven by one coherent idea. The only person who could communicate that was the CEO—me.

These communications were absolutely critical to me in the early days. My message was quite simple. I stood before IBM employees all over the world, looked into their faces and said, "Clearly, what we have been doing isn't working. We lost $16 billion in three years. Since 1985, more than 175,000 employees have lost their jobs. The media and our competitors are calling us a dinosaur. Our customers are unhappy and angry. We are not growing like our competitors. Don't you agree that something is wrong and that we should try something else?"

I also discovered the power of IBM's internal messaging system, and so I began to send employees "Dear Colleague" letters. They were a very important part of my management system at IBM. I sent the first one six days after I'd arrived:

April 6, 1993

Office of the Chairman

MEMORANDUM TO: All IBM Colleagues

SUBJECT: Our Company

It wasn't long after I arrived that I discovered on my office PC that PROFS mail is an important vehicle of communication within IBM. Thanks to all who sent greetings, best wishes, suggestions, and advice.

I'm sure you understand that I cannot reply to every message. But I did want to take this early opportunity to acknowledge some frequent, serious themes in your correspondence.

I was moved by your intense loyalty to IBM and your very clear desire to restore IBM—as quickly as possible—to market leadership. This has been as true of those leaving the company as of those staying here. It all underscores that our strength is indeed our people and their commitment to success.

Some of you were hurt and angered by being declared "surplus" after years of loyalty, and by some reports in the press about performance ratings.

I am acutely aware that I arrived at a painful time when there is a lot of downsizing. I know it is painful for everyone, but we all know, too, that it is necessary. I can only assure you that I will do everything I can to get this painful period behind us as quickly as possible, so that we can begin looking to our future and to building our business.

I want you to know that I do not believe that those who are leaving IBM are in any way less important, less qualified, or that they made fewer contributions than others. Rather, we ALL owe those who are leaving an enormous debt of gratitude and appreciation for their contributions to IBM.

Finally, you've told me that restoring morale is important to any business plans we develop. I couldn't agree more. Over the next few months, I plan to visit as many of our operations and offices as I can. And whenever possible, I plan to meet with many of you to talk about how together we can strengthen the company.

Lou Gerstner

The reaction from IBM employees was overwhelmingly positive and, for me during the dark, early days, a source of comfort, support, and energy. Said one:

Tears of joy came to my eyes.

Another wrote to me, simply:

Thank you, thank you, thank you. Sanity is returning to IBM.

At the same time, IBM employees were never afraid to speak their minds when it came to expressing feelings of opposition. I got e-mail messages so frank, so candid, so blunt—well, I'll just say that when I was younger, I would never have sent such messages to my boss, much less the CEO. One employee wrote to me:

> GIMME A BREAK. Do some real work. Cut the order cycle time. Get the new products on the market. Find new markets. Listen to the folks that are not our current customers but would be if we had products for them.
>
> Stop this bleeding heart stuff. Do things that will keep you from having to trash more and more people every 6 months.

Another greeted my arrival this way:

> Welcome and don't worry about not knowing very much about microchips, just as long as you don't get them mixed up with chocolate chips.

One employee, even as his employer was burning and sinking to the delight of our competitors, had the time and inclination to critique my entire visit to an IBM facility:

> There were three areas in which I thought your attitudes and perspectives could be healthier. You come across as so accessible and willing to accept feedback that I feel comfortable sharing them with you in a note.
>
> 1—You gave a pecking order of importance for IBMers; first, the Customer, second IBM, third one's own unit. This sounds like a McKinsey hierarchy. I submit a more appropriate number one on the list, and an IBM tradition, is one's self—the rest of the list

could stay the same. Respect for the individual is fundamental to health, whether it be the health of an individual, of an organization, or of a society. (The McKinsey hierarchy, in which the individual is somewhere after Customer and company, burns out employees and their families.)

You described the need for us to examine ourselves and the way we've been doing business. I also value self-reflection and suggest the following as areas in which you may benefit from introspection. (These are opportunities for you to lead by example.)

2—You seemed to want to compete and placed a great deal of importance on beating the competition. I recognize this attitude is culturally endorsed, but I also believe it is unnecessary, unhealthy, and less productive than other forms of social interactions. For example, the competitive mindset within IBM (IBMers beating IBMers) is something you railed against. You also emphasized the need to delight Customers. I agree with that as a goal and submit that it is a different goal from "beating the competition." Processes we craft to reach these goals would be different. If we aren't clear about our goal, we will most likely fail to develop sound processes to achieve it.

A couple of particulars in this area. You mentioned "beating the stuffing" out of someone and "ripping off their face." Don't these sound like unhealthy attitudes? These "someone's" are people with friends and families. They may even be your friends and relatives. Competition, a structure and attitude in which involved parties try to prevent each other from reaching their goal is, at its very core, disrespectful of the individual.

I had sent you an audiotape on this topic ("Cultural Heresy: The Case Against Competition") and a short description of the tape. Apparently the tape and letter were intercepted by an administrative assistant and never reached you. If you are interested in exploring this topic further, I can resend you the tape.

You claimed that the most important measurement of our

success was the percentage of the information technology budget we had for each Customer. This seems to me to be a case of limited thinking. A percentage is finite and can never get larger than 100. Using this metric, any gains one company makes has to be at the expense of one or more other companies. If we think expansively, if we think about how we can make the pie bigger, then everyone could experience a win. For instance, if there is more money spent on information technology because of its increasing value, we could be losing percentage points while growing and making more money. (I suspect we were losing percentage points in the early '80s when we were expanding and making $1 billion a quarter.) Conversely, how interested are we in achieving 100% of the market for card readers?

Though I've focused on the "areas of improvement" in this note, I want to re-emphasize that I admire and respect you for the job you've already done and are doing. I'm looking forward to working with you.

[Name deleted]

P.S.—I don't know if it's true, but I heard that in preparation for your visit to the (Raleigh, North Carolina) site, the route you would take was planned and the halls you would walk down or see had their walls painted and new carpeting laid. I was wondering if you knew whether or not this was true and if it was true, what you thought about it.

Sometimes I had to bite my tongue—almost in half. All I can say is, it was a good thing for some people that I was too busy to reply to all my e-mail![1]

---

[1]   Appendix A contains additional examples of the Dear Colleague memos that were so important to our transformation.

Creating a Global Enterprise

What we had done thus far was to put out the fire. Now we needed to rebuild the fundamental strategy of the company. That strategy, as I had been saying for six months, was going to revolve around my belief that the unique opportunity for IBM—our distinctive competence—was an ability to integrate all the parts for our customers.

However, before I could integrate for our customers, I first had to integrate IBM! So, as our strategy people worked on fleshing out short- and long-term plans, I turned my attention to three areas that, if not fundamentally changed, would disable any hope of a strategy built around integration: organization, brand image, and compensation.

Remaking the Organization

IBM is arguably the most complex organization anywhere in the world outside government. It is not just its sheer size ($86 billion in

2001 sales), nor its far-flung reach (operating in 160-plus countries). What drives IBM's unique complexity is twofold. First, every institution and almost every individual is an actual or potential customer of IBM. In my previous occupations, we could always identify a dozen or so key customers in one or two industries that really defined the marketplace. Not so at IBM. We had to be prepared to serve every institution, every industry, every type of government, large and small, around the globe.

The second complexity factor is the rate and pace of the underlying technology. Again, in prior incarnations, my management team and I could identify four or five companies or organizations that had been our competitors for the past twenty years and would probably continue to be our competitors for the next twenty. In the information technology industry, literally thousands of new competitors sprang up every year—some in garages, some in universities, some in the hearts and minds of brilliant entrepreneurs. Product cycles that used to run for ten years dwindled to nine or ten months. New scientific discoveries overwhelmed planning and economic assumptions on a regular basis.

It is not surprising, therefore, that in the face of this large global span and uniquely diverse set of customers and an ever-changing technological base, organizing IBM was a constant challenge.

One other factor made it particularly interesting—the nature of the IBM employee base. We are not a company of management and workers. We are a company of 300,000-plus professionals, all of whom are bright, inquisitive, and (alas) opinionated. Everybody had his or her view of what the first priority should be and who should manage it.

As IBM grappled with this recipe for cacophony, the company evolved over the years in two directions: powerful geographic units that dealt with IBM's global reach, and powerful product divisions that dealt with the underlying technological forces. Missing from this structure was a customer view. The geographic regions, for the most part, protected their turf and attempted to own everything that went on in their region. The technological divisions dealt with what they thought could be built, or what they wanted to build, with little concern about customer needs or priorities.

I had experienced this firsthand at American Express and was determined to see it changed soon. There had been eleven different currencies in which the American Express Card was issued when I arrived; there were more than twenty-nine when I left. As we moved the Card around the world, we needed common systems from IBM, our primary information technology vendor, and we needed support in every major country in the world.

I was always flabbergasted to find that when we arrived in a new country (Malaysia or Singapore or Spain), we had to reestablish our credentials with the local IBM management. The fact that American Express was one of IBM's largest customers in the United States bore no value to IBM management in other countries. We had to start over each time, and their focus was on their own country profit and loss, not on any sense of IBM's global relationship with American Express.

The same was true of products. Products used in the United States were not necessarily available in other parts of the world. It was enormously frustrating, but IBM seemed to be incapable of taking a global customer view or a technology view driven by customer requirements.

One of my first priorities was to shift the fundamental power bases inside IBM. In the United States alone, there was a national headquarters, eight regional headquarters, multiple area headquarters under the regions, and, finally, local units called "trading areas." Each was run by a profit center boss who sought aggressively to increase his or her own resources and profits. Say a banking client in Atlanta wanted a solution involving retail banking. Never mind that the best banking experts were in New York City or Chicago. The local boss would often ignore those resources and use his or her own people. (One day I looked at the financial newswires and was shocked to see that IBM's trading area for Alabama and Mississippi had sent to the media its own earnings news release.)

Staff units abounded at every level. Outside the United States, the structure was even more rigid, like in Europe, with its 23,000 support people. Other IBMers practically had to ask permission to enter the territory of a country manager. Each country had its own independent system. In Europe alone we had 142 different financial systems. Customer data could not be tracked across the company. Employees belonged to their geography first, while IBM took a distant second place.

Breaking Up the Fiefdoms

I declared war on the geographic fiefdoms. I decided we would organize the company around global industry teams. I had first encountered the power of this kind of structure when I had been a very young consultant at McKinsey. We'd conducted a seminal organization study for what was then Citibank. The result was to transform Citibank from a geographical organization to a global, customer-oriented organization, and it became the model for most financial institutions over the next decade.

With this model in mind, I asked Ned Lautenbach, then head of all non–United States sales organizations, to build a customer-oriented organization. It was a painful and sometimes tumultuous process to get the organization to embrace this new direction, but by mid-1995 we were ready to implement it. We broke our customer base into twelve groups: eleven industries (such as banking, government, insurance, distribution, and manufacturing) and a final category covering small- and medium-size businesses. We assigned all of the accounts to these industry groups and announced that the groups would be in charge of all budgets and personnel. The response from the country managers was swift and predictable: "It will never work." And: "You will destroy the company!"

I'll never forget one run-in with the head of the powerful Europe, Middle East, and Africa unit. During a visit to Europe I discovered, by accident, that European employees were not receiving all of my company-wide e-mails. After some investigation, we found that the head of Europe was intercepting messages at the central messaging node. When asked why, he replied simply, "These messages were inappropriate for *my* employees." And: "They were hard to translate."

I summoned him to Armonk the next day. I explained that he had no employees, that all employees belonged to IBM, and that from that day on he would never interfere with messages sent from my office. He grimaced, nodded, and sulked as he walked out the door. He never did adapt to the new global organization, and a few months later he left the company.

Although we implemented the new industry structure in mid-1995, it was never fully accepted until at least three years later. Regional heads clung to the old system, sometimes out of mutiny, but more often out of tradition.

We needed to do a massive shift of resources, systems, and processes to make the new system work. Building an organizational plan was easy. It took three years of hard work to implement the plan, and implement it well.

I'll never forget one particularly stubborn—and inventive—country general manager in Europe. He simply refused to recognize that the vast majority of the people in his country had been reassigned to specialized units reporting to global leaders.

Anytime one of these new worldwide leaders would pay a visit to meet with his or her new team, the country general manager, or GM, would round up a group loyal to the GM, herd them into a room, and tell them, "Okay, today you're database specialists. Go talk about databases." Or for the next visit: "Today you're experts on the insurance industry." We eventually caught on and ended the charade.

[9]

Reviving the Brand

All of our efforts to save IBM—through right-sizing and reengineering and creating strategy and boosting morale and all the rest—would have been for naught if, while we were hard at work on the other things, the IBM brand fell apart. I have always believed a successful company must have a customer/marketplace orientation and a strong marketing organization. That's why my second step in creating a global enterprise had to be to fix and focus IBM's marketing efforts.

IBM won numerous awards in the 1980s for its ingenious Charlie Chaplin commercials, which had introduced the IBM personal computer. By the early 1990s, however, the company's advertising system had fallen into a state of chaos. As part of the drive toward decentralization, it seemed that every product manager in just about every part of the company was hiring his or her own advertising agency. IBM had more than seventy ad agencies in 1993, each working on its own, without any central coordination. It was like seventy tiny trumpets all tooting simultaneously for attention. A single issue of an industry trade magazine could have up to eighteen different IBM ads, with eighteen different designs, messages, and even logos.

In June 1993 I hired Abby Kohnstamm as the head of Corporate

Marketing for IBM. She had worked with me for many years at American Express. What we had to do here was so important and urgent that I wanted someone who knew me and how I managed, and with whom I could speak in shorthand.

Abby had an especially tough challenge. There had never been a true head of marketing in IBM. Few people in the business units understood or accepted her role, and at first they tried to ignore her. IBM was built on technology and sales. And, in IBM at that time, the term "marketing" really meant "sales." Using the broadest definition, sales is about fulfilling the demand that marketing generates. When it's done well, marketing is a multi-disciplinary function that involves market segmentation and analysis of both competitors and customer preferences, corporate and product brand management, advertising, and direct mail. That's only a partial listing. While IBM clearly had to sell more of what it made, it also had to recast its image and reestablish its relevance to the marketplace. When I arrived at IBM, marketing was not considered a distinct professional discipline, and it was not being managed as such. I told Abby to take sixty days to do a situation analysis.

Her research found that despite our well-chronicled problems, the overall IBM brand was still strong. Customers believed that if they bought an IBM product, it would be a good one. As I had expected, our biggest strength was as a unified brand, and not as each of our parts. Consequently, the marketing mission would be to articulate why customers would want to do business with an integrated IBM.

Abby knew she had to end the dissonance. We got there in stages because, while you can force anything down the throat of an organization, if people don't buy into the logic, the change won't stick. Stage one was weaning IBM executives off the luxury of having their own advertising budgets, their personal agencies, and the discretion to order up an ad anytime they wanted to do so. One month there'd be no IBM advertising in important industry magazines; the next month we'd have so many pages that it seemed as if we were sponsoring a special issue. The latter was especially true in November and December, when marketing departments wanted to spend leftover dollars in their budgets.

Abby's job was to get control of the spending and the messages. I asked her to present a plan to the newly formed Worldwide Management Council at our conference center in Palisades, New York. It was a tough meeting, but she did a very smart thing. When the thirty-five WMC members walked into the room, they found every wall adorned with the advertising, packaging, and marketing collateral of all our agencies. It was a train wreck of brand and product positioning.

After her presentation, I posed one question: "Does anyone doubt we can do this better?" There was no discussion.

One Voice, One Agency

Abby decided to consolidate all of IBM's advertising relationships into a single agency—not just in the United States, but around the world. At the time, it was the largest advertising consolidation in history. Few people knew about her plan at first—a handful of people inside IBM, and only the chief executives of the agencies under consideration. There was no formal advertising review process. No creative development. No presentations. Abby narrowed down the list to four agencies, including just one that was then handling any of the IBM accounts. Over four weeks, she held a series of two-day meetings in hotels (with people on both sides using aliases) to gauge chemistry, thinking, and how each would approach a challenge this big.

She and her small team of IBMers unanimously settled on Ogilvy & Mather, which had solid worldwide expertise and experience. That was exactly what IBM needed, since the agency would manage advertising for all of our products and services, as well as our overall brand, around the world.

Before we signed off on the deal, I asked Abby to bring the top three people at O&M to our corporate headquarters. We were about to bet the IBM brand on these people, so I wanted to make sure we were all clear about the stakes we were playing for—to look them in the eye and hear them commit to the success of this effort, no matter what it was going to take. Interestingly, the meeting revealed that they had a parallel objective. They were betting a big part of their future on IBM—and resigning from several of their existing accounts. So, in fact, we were looking each other in the eye.

Abby had my complete support, but others were a tougher sell, both inside and outside the company. Many of the product and geographic units adopted a "this too shall pass" approach—up until the time when we centralized most of the advertising spending and the media buying and went to global contracts. The ad community itself was in absolute shock. Not only was this not done in the advertising world—but by stodgy, trouble-plagued IBM?

The New York Times covered the consolidation on page one. *Advertising Age* called it the "marketing shot heard 'round the world." But it was a mostly positive shot. *Ad Age* went on to say: "Because computer products, brands, and publications have few geographic boundaries, a world approach makes sense. . . . A single agency meshed neatly with Mr. Gerstner's strategy to centralize controls and bring independent units like the PC division back into the fold."

Always critical, *The Wall Street Journal* called the decision "audacious" and "fraught with risk." The paper warned: "If the agency doesn't devise an instantly winning campaign, it could set IBM's recovery back for months."

Far from it. Despite the fierce kicking and screaming of many local managers, the first campaign debuted in 1994 under the theme "Solutions for a Small Planet." The innovative TV spots—featuring an international cast, from Czech nuns to old Parisians speaking in their native language with all the dialogue subtitled—were highly acclaimed.

The campaign reaffirmed important messages: IBM was global,

and we were staying together as a world-class integrator. At the same time, it signaled that we were a very different company—able to change and make bold decisions, just as we had done with the decision to consolidate; able to move quickly; able to take risks and do innovative things; and we were more accessible. The campaign humanized our brand.

In conjunction with the creative work, we completely overhauled our budgeting and media buying. We knew we'd save money through the consolidation, and we certainly did. But that wasn't the reason to do it. In fact, we immediately doubled our investment in marketing and advertising, and we've sustained that level of investment over the years.

"Solutions for a Small Planet" was followed by a campaign that coined the term "e-business" and helped establish IBM as the leader of the most important trend in the industry at that time (more on that later).

Against all odds, Abby Kohnstamm had made something great and impactful happen. She had to build the marketing function from scratch and simultaneously create a unified, global campaign for a company that had a recent history of dysfunctional, fractious, competing messaging. All of this rubbed mightily against the historical culture of IBM. Abby was another hero of the turnaround.

Resetting the Corporate Compensation Philosophy

The "old" IBM had very fixed views about compensation; much of it, I suspect, had been derived from the management philosophy of Tom Watson, Jr., the man who had created the great IBM of the 1960s and 1970s. Since the company's performance during that time had been so extraordinary, it would be foolish to say it was not an effective compensation system.

Let me briefly describe the system I discovered when I arrived.

First, compensation at all levels consisted predominantly of salary. Relatively little was paid in bonus, stock options, or performance units.

Second, there was little differentiation in the system.

- Annual increases were typically given to all employees except those rated unsatisfactory.

- There was very little variance in the size of the annual increase between a top-ranked and a lower-ranked employee.

- Increase sizes were in a small band around that year's average. For example, if there was a 5 percent increase in budget, actual increases fell between 4 percent and 6 percent.

- All employee skill groups (such as software engineers, hardware engineers, salesmen, and finance professionals) were paid the same within a salary grade level, regardless of the fact that some skills were in higher demand externally.

Third, there was a heavy emphasis on benefits. IBM was a very paternal organization and provided generously for all forms of employee support. Pensions, medical benefits, employee country clubs, a commitment to lifelong employment, outstanding educational opportunities—all were among the best of any United States company.

From what I can tell, there was little benchmarking of IBM's practices vis-à-vis other companies. In a sense IBM *was* the benchmark and decided on its own what it wanted to do.

Basically it was a family-oriented, protective environment where equality and sharing were valued over performance-driven differentiation.

I was well aware of the strong commitment IBM held for its employees long before I joined the company. However, as good as it might have been during IBM's heyday, the old system was collapsing amid the financial crisis that preceded my arrival. Tens of thousands of people had been laid off by my predecessor—an action that shocked the very soul of the IBM culture. The year before I arrived, limits were put in place on future medical benefits, setting the stage for cost sharing by employees and retirees—another very difficult break with the past for IBMers.

The old system was not only out of touch with the realities of the marketplace, but it was unable to satisfy the paternalistic underpinnings of the historical IBM culture. Consequently, it made fixing the company very difficult and made employees sad and cynical. We needed a whole new approach—and we needed it fast.

Pay for Performance

We made four major changes to our compensation system, and I'll describe them in a moment. Behind all of them was a fundamentally different philosophy than what had been followed in the past, best described in this list:

OLD	NEW
Commonality	Differentiation
Fixed rewards	Variable rewards
Internal benchmarks	External benchmarks
Entitlement	Performance

This was all about pay for performance, not loyalty or tenure. It was all about differentiation: Differentiate our overall pay based on the marketplace; differentiate our increases based on individual performance and pay in the marketplace; differentiate our bonuses based on business performance and individual contributions; and differentiate our stock-option awards based on the critical skills of the individual and our risk of loss to competition.

Let me comment now on a few of the specific changes we made.

Stock Ownership

Have you ever wondered where the Watson family fortune is? Certainly Mr. Watson, Sr., who started at the company in 1914, and his son, who was CEO during the great growth phase of the company (with both of these tenures added together, they ran the company for fifty-six consecutive years), had the opportunity to amass a net worth on the order of the Fords, Hewletts, and Waltons. Surely there could have been a Watson Foundation as powerful as the Ford Foundation or the Hewlett Foundation. But there was no such aggregation of wealth!

Why? It appears that both Watsons had strong views that limited their ownership of IBM stock. Tom, Sr., never owned more than 5 percent of the company and refused to grant stock options to himself or other executives. He liked cash compensation and was paid a salary plus a percentage of the profits of the company.

Tom, Jr., started a stock-option program in 1956, but it was limited to a very few executives. Regarding his own ownership, in his book, *Father, Son, and Co.,* he stated that he stopped taking options in 1958 (he was CEO until 1971), believing that his $2 million worth of options at that time would be worth tens of millions of dollars in the future. Apparently he felt that was enough.

It appears that for Tom Watson, Jr., stock options were intended solely to reward executives—and not to link executives to the company's shareholders. In fact, in the aforementioned book, he stated that "the model corporation of the future should be largely owned by the people who work for it, not by banks or mutual funds, or shareholders who might have inherited the stock from their parents and done nothing to earn it."

While I think Tom Watson and I share a lot of common beliefs (in particular, our passion for winning), here we part ways.

I wanted IBMers to think and act like long-term shareholders— to feel the pressure from the marketplace to deploy assets and forge strategies that create competitive advantage. The market, over time, represents a brutally honest evaluator of relative performance, and what I needed was a strong incentive for IBMers to look at their company from the outside in. In the past, IBM was both the employer and the scorekeeper in the game. I needed my new colleagues to accept the fact that external forces—the stock market, competition, the changing demands of customers—had to drive our agenda, not the wishes and whims of our team.

However, beyond their role as a connector to the outside world, stock options played an even more important role in my early days at IBM. I had made the decision to keep IBM together. Now I had to make that decision pay off. I just told you about the role that organization and branding decisions played in supporting this integration strategy. Nothing, however, was more important to fostering a one-for-all team environment than a common incentive compensation opportunity for large numbers of IBMers—an opportunity that was heavily dependent on how the overall corporation performed. I repeatedly told my team that we don't report software profits per share or PC profits per share—only IBM consolidated profits per share. There was only one financial scoreboard, and it was the stock price reported every day in the media. People had to understand that we all benefited when IBM as a whole did well and, more often than not, lost out when we functioned as a disjointed operation.

Consequently, we made three big changes to the IBM Stock Options Program. First, stock options were offered to tens of thousands of IBMers for the first time. In 1992, only 1,300 IBMers (almost all high-level executives) received stock options. Nine years later, 72,500 IBMers had received options, and the number of shares going to nonexecutives was two times the amount executives received.

I want to emphasize that the decision to make options widely available to employees is not a general tenet of my personal management philosophy. In fact, I am not a fan of corporate plans that promise a minimal grant of options to every employee in a company. Most employees view such options as nothing but a delayed form of salary. As soon as they can, they cash them in.

However, IBM is different—perhaps from any other company in the world. As I said earlier, it has, for the most part, a single class of knowledge workers. Second, it does not have multiple businesses. It has one gigantic, $86 billion business.

Engineers, marketers, designers, and other employees around the globe had to act in sync if we were going to pull off the integration of IBM. I had to have all these people thinking as one cohesive unit, and granting stock options to thousands of them would help focus attention on a common goal, a common scorecard of performance. I needed to convince IBMers they were better off working as a singular enterprise— one team and not separate fiefdoms. If I could not do that, my entire strategy for turning around the company would fail.

The second decision regarding stock options involved executives, and it was far more straightforward: We made stock-based compensation the largest element of executives' pay, downplaying annual cash compensation relative to stock appreciation potential. This is part of my management philosophy. Executives should know they don't accumulate wealth unless the long-term shareholders do the same.

The third, and final, decision regarding options was also based on a view I hold very strongly. Executives at IBM were not going to be granted stock options unless they concurrently put their own money into direct ownership of company stock. We established guidelines that effectively said: "You have to have some skin in the game." No free ride.

EXECUTIVE STOCK OWNERSHIP GUIDELINES

The value, in U.S. dollars, of the IBM stock you are expected to own is determined by your position and a multiple of your combined annual base salary and annual incentive at target:

POSITION	MINIMUM MULTIPLE
Chief Executive Officer	4
Senior Vice President	3
Other Worldwide Management Council Member	2
Other Senior Leadership Group Member	1

Every executive had to be in the same position as a shareholder: stock up, we'll feel good; stock down, we'll feel pain (real pain—not the loss of a theoretical option gain). I bought stock repeatedly on the open market in the early days, because I felt it was important to have my own money at risk.

A Distasteful Necessity

One historical footnote on the subject of stock options: In 1993 I used options in a way I find personally distasteful, but which was necessary because of our financial crisis. I realized that I had no holding power over key technical and managerial talent at IBM and that our competitors were systematically raiding us to grab our best people. I knew that at the time it was important to reduce our overall number of employees, but in a crisis it was even more important to retain our most promising people.

I looked around for choices, one of which would have been to introduce new stock options for key people. However, no shares were available under the existing stock-option plan, and the only way to get them would have been to call a special shareholders' meeting. After the contentious Tampa shareholders' meeting, imagine my asking IBM shareholders for yet another meeting solely to approve more stock for management!

I decided to offer the people we most wanted to retain an opportunity to turn in their basically worthless existing options for new, lower-priced options. I loathe doing this because rewriting the rules halfway through is not the way to play the game, but I was able to overcome my personal bias by setting very specific terms that I believe, in light of the dire circumstances, made this a viable and appropriate program for shareholders at the company at that time. And we excluded senior executives. They had played a role in creating our problems. They had to keep their old options at the old prices and work to solve the problems.

This was a very important and successful program. I can't give specific numbers, but I know it helped to retain a lot of important people who had been tempted to join competitors but are now in leadership positions at IBM. Also, it sent a message to everyone—including the executives who were excluded—that we really intended to tie our performance to share price and that we wanted to align our interests directly with those of the shareholders. And, finally, it sent a message, the first of many to follow, that compensation at IBM was going to be performance-based, not simply length-of-employment-based.

Other Changes

I've described in detail the changes made to the stock-option program principally because I wanted to underscore my belief that you can't transform institutions if the incentive programs are not aligned with your new strategy. I'll conclude this chapter with a description of several other changes we made to bring the compensation system in line with the new IBM.

Prior to my arrival, bonuses were paid to executives based solely on the performance of their individual units. In other words, if your operation did well but the overall corporation did poorly, it didn't matter. You still got a good bonus. This encouraged a me-centered culture that ran counter to what I was trying to create at IBM.

Therefore, beginning in 1994, we instituted a huge change. All executives would have some portion of their annual bonus determined by IBM's overall performance. The most unusual part of this plan involved the people who reported directly to me—the highest-level executives, including those who ran all our business units. From then on, their bonuses were to be based *entirely* on the company's overall performance. In other words, the person running the Services

Group or the Hardware Group, had his or her bonus determined not by how well the unit performed, but by IBM's consolidated results. Executives at the next level down were paid 60 percent based on consolidated IBM results, 40 percent on their business unit results. The system cascaded down from there.

Of all the changes I made in 1993 and 1994, nothing else had the impact that this move had in sending a message throughout the company: "We need to work together as a team. Gerstner's not kidding. He really wants us to make integration the centerpiece of our new strategy."

We made a similarly bold statement to our employees. In the mid-1990s we introduced "variable pay" globally across IBM. This was our way of saying to all IBMers that if the company could pull off its turnaround, each and every one of them would share in the rewards. Over the next six years, $9.7 billion was paid out to IBMers worldwide (with a few exceptions in countries where this program was not permitted by law).

The variable pay amounts were also tied directly to overall IBM performance to ensure that everybody knew that if they worked hard at collaboration with colleagues, doing so would pay off for them.

The final change we made was the least strategic but the most controversial: paring back the paternalistic benefits structure. We did not undertake these changes because we thought the highly generous support system was bad per se. Believe me, I would have loved to continue the employee country clubs and the no-cost medical plans. We cut back on these plans because the company could no longer afford the level of benefits. The high profit margins of the 1970s and 1980s were gone—forever. We were fighting for our lives. None of our competitors offered anything close to the IBM benefits package. (Even now, after all the changes we made, IBM benefits programs are among the most generous of any United States–based multinational corporation.)

Also, we changed benefits because the old system was geared to the company's prior commitment to lifelong employment—for example, the bulk of pension benefits accrued after thirty years of service.

The new IBM was not a place where jobs could be guaranteed for life (nor was the old IBM after it got in trouble). So we had to create benefits programs that were more appropriate to a modern workforce.

Some of these benefits changes created a great furor among a small group of IBM employees, but the vast majority of IBMers unselfishly understood that the changes were absolutely necessary for the company to survive and grow. More important, IBMers at all levels embraced the overall shift in compensation philosophy—fewer paternal benefits, but a far larger opportunity for everyone to participate in the rewards of our success through variable pay programs, stock-purchase and -option plans, and performance-based salary increases.

Postscript: The First Year Ends

The first year ended on a very sad note: the death of Tom Watson, Jr., in December 1993. I had seen Tom only one more time after he had ridden to work with me that morning in April. He expressed his delight that I had decided to keep the company (he again called it "my company") together.

As I sat in my pew at his memorial service, I couldn't help but wonder what he might have thought about the massive changes we had made in only nine months, and the reactions—both positive and negative—to these changes expressed by employees and outsiders.

I found myself wishing that Tom and I had had a chance to have lunch or dinner to talk about the "new Blue" that was beginning to emerge but still had so very far to go. I felt strongly that like most other great people who built great things, Tom Watson was at heart an agent of change.

Exhausted but encouraged, I flew to Florida for my annual Christmas vacation on the beach. I had a lot to think about.

[11]

Back on the Beach

On a gray morning ten months after I had taken that walk on the beach in Florida, thinking about my conversations with Jim Burke and deciding whether or not to parachute into IBM, I found myself back on the same beach, mulling over the extraordinary events that had transpired since that time.

I had to admit, I felt pretty good. Few had given us any chance of saving IBM, but I knew now that the company was going to make it. We'd stopped the bleeding, reversed the breakup plan, and clarified IBM's basic mission. The holes in the hull had been patched. This ship was not going to sink.

My thoughts turned to what lay ahead. What would Act II look like? Logic and my own experience dictated a straightforward set of priorities: Invest in new sources of growth, build a strong cash position, and do a more rigorous assessment of our competitive position.

However, doing all of that wasn't enough. Even if we restored growth, even if we built up some momentum with customers, and even if we made the company more efficient and less bureaucratic—that wouldn't truly bring IBM back. For IBM's turnaround to be successful, this company would have to regain its former position of leadership in the computer industry and in the broader world of business.

I can't remember whether I smiled, laughed, or shook my head. But that question—Could IBM lead again?—gave me pause. Certainly it would be easy, and expected, for the CEO to declare that the company would lead again. But as I thought about what it would actually take for that to happen, all of my original doubts about accepting this job came flooding back.

First of all, the track record of IT companies that had been pulled back from the brink was dismal. I thought of Wang, Data General, Sperry-Burroughs (today's Unisys), DEC. Even when companies were rescued, they usually survived as also-rans or found another partner to merge with or put themselves up for sale.

Most troubling was the computer industry's trajectory. Basically, it was moving away from IBM's traditional strengths. The PC wasn't an endgame, and the mainframe wasn't dead. But it was obvious that in the emerging computing model—of which the PC had been a harbinger—power was migrating rapidly away from centralized computing systems and traditional IT. And this was, in turn, changing the mix of IT customers. IBM sold to large businesses, governments, and other institutions. But more and more, IT was being bought by consumers, small businesses, and department heads inside big companies.

The emerging computer model was also changing what those customers were buying. We built industrial-strength, behind-the-scenes computers and software while the world, it seemed, was moving to desktop, laptop, and palmtop. All of our research and development, engineering, and rigorous testing ensured that our product never went down. Reliability, dependability, and security— these were the bedrock of IBM's brand. But people didn't seem to mind rebooting their PCs three times a day.

We sold through a direct sales force—the vaunted, blue-suited IBM customer representative; part salesperson, part business and technology consultant. A tremendous asset, but also the most expensive way to sell any product or service. The market was abandoning that model and going to retailers and toll-free numbers.

Overall, the tasks we were being asked to take on were spreading beyond the domain of the CIO and into every corner of business operations—places where IBM had not, in general, ventured and where we lacked strong customer relationships. And, most ominously, value and profit margins were shifting away from hardware, which was becoming more commoditized, and toward software (and, it was beginning to appear, toward services).

Then, look at who we were up against! The people running our competition were, without doubt, the next generation of hyper-capitalists: Bill Gates, Steve Jobs, Larry Ellison, and Scott McNealy. These guys were hungry, and they stayed hungry no matter how much wealth they accumulated. And it was awe-inspiring the way they ran their companies, the people they attracted, how they paid them, their work ethic—young, aggressive, flexible, willing to work around the clock. The whole Silicon Valley ethos—lightning speed to market with just-good-enough products—wasn't simply foreign to IBM, it was an entirely new game.

Even without considering this formidable competition, our own strategy raised some daunting implications. What we had done so far to unify IBM—reorganizing around industries rather than countries, consolidating our marketing, and changing our compensation plans—had been relatively easy to accomplish. What lay ahead—devising a strategy for a fundamentally new world and reinventing an encrusted culture from the DNA out—that was a challenge of a vastly different order.

I asked myself: "How did I get into this? Is it an impossible task?" It would be hard to say no with a straight face, even to my closest colleagues. I could see clearly what the remaining four years of my contract would look like. We had a chance to grow. Maybe we could take on and displace a few competitors in some segments. But lead the industry? That mountain looked too high to climb. And if I set that task as the goal, I stood a very good chance of failing—very visibly.

I walked more, thought more, and the clouds began to clear. Yes, those were daunting obstacles, but wasn't that why I'd come here? Didn't this make the challenge that much more intriguing?

And wasn't it worth the risk? If we didn't aspire to leadership, one thing was clear: The company would never really come together, never really achieve its potential. And that would be a shame.

I recalled the comment Jim Burke had made about IBM's being a national treasure. In fact, Burke was just one of several people who had expressed that phrase to me. Shortly after the announcement of my appointment, I had run into Joshua Lederberg, a research geneticist and Nobel laureate on the street in Manhattan. I knew him from the board of Memorial Sloan-Kettering Cancer Center. "You're going to IBM," he'd said to me. I'd said I was. "It's a national treasure," he'd said. "Don't screw it up."

At the time, I thought this reverence was a bit over the top. Up to that point in my career, I had dealt mostly with businesspeople who were motivated by basically two things: money and power. And we were in the midst of one of the most swashbuckling eras in the history of commerce. In contrast, when I got to IBM, I felt as though I had entered a time warp and gone back to the 1950s.

The fact was, IBM had grown on me. I had come to understand what Jim Burke and Dr. Lederberg meant. The corporation *was* important—not only in what it did for customers and governments and universities, and not only for what it invented, impressive and meaningful as those accomplishments were. It was also important for the kind of corporate behavior to which it aspired. IBMers were battered, bruised, and confused. Many had retreated into a self-protective shell. But underneath that, they were still motivated by a genuine love of their company and of doing the right thing.

You could make fun of IBM all you liked. (Our competitors certainly did.) But for issues that really mattered—when it was a question of national defense, or our children's health, or serious scientific discovery—IBM was essential. Forgive my hyperbole, but in an industry increasingly run by mad scientists and pied pipers, we *needed* to succeed.

I hadn't left consulting and gone into management just to be Mr. Fix-It or simply for the pleasure of getting into the game. More than anything, I like to win. But the issue here went beyond winning. For the first time in my career, I was in a position to make a different kind of mark, to help something truly important live and thrive. I wasn't going to walk away from that. As I headed back to our family's beach house, I began to feel intense excitement. I told myself: "You know, we could actually pull this off!"

So the die was cast. I'd concluded it was insufficient—for me personally and for the institution—to call it a day after battling back from IBM's near-death experience. We were going to take our best shot at making the long climb back to industry leadership.

Heading into IBM, I would have bet large sums of money that these frenetic early months of decision making and action taking to stabilize the patient would be the hardest work of my professional career. I would have bet wrong. It had been difficult, even painful. The issues, however, were reasonably obvious, the problems were easy to parse, and the remedial actions were straightforward.

Now, after nearly a decade of subsequent work—and with the benefit of a little distance from the day-to-day existence of the CEO—I can say without hesitation that what came next was far more difficult. If nothing else, those first twelve to eighteen months at least had the benefit of adrenaline-stoked intensity—many highs, an equal number of lows, but never time to celebrate one or dwell on the other because we were literally in a situation in which every minute counted.

What I'd come to realize during this second walk on the beach was that after all that initial work had been completed, we'd gotten ourselves only to the starting line. The sprint was over. Our marathon was about to begin.

While the issue was no longer as stark as the demise or survival of IBM, the ultimate fate of this "national treasure" was far from settled. What would happen through the second half of the 1990s would determine whether IBM was merely going to be one more pleasant, safe, comfortable—but fairly innocuous—participant in the information technology industry, or whether we were once again going to be a company that mattered.

The outcome of that race has been abundantly documented. By 1997 we'd declared the IBM turnaround complete. Inside the company we were talking openly about getting back on top and once again setting the agenda for our industry—aspirations that, when we began, would have seemed excessively ambitious at best, delusional at worst.

Before I stepped away in March 2002, we were number one in the world in IT services, hardware, enterprise software (excluding PCs), and custom-designed, high-performance computer chips (see Chapter 16). The IBM team had staged comebacks in multiple markets where we'd previously been getting sand kicked in our faces. We'd revamped and reinvigorated traditional product lines, launched new growth businesses, and jettisoned several others that were vestiges of the earlier era.

At a higher level, we had articulated and then led the future direction of the industry—a future in which business and technology would not be separate tracks but intertwined; and a future in which the industry—in a remarkable about-face—would be driven by services, rather than hardware or software products. We'd coined the term "e-business" and played a leadership role in defining what was going to matter—and what wasn't—in a networked world.

The IBM workforce increased in size by about 100,000 people. Our stock split twice and increased in value by 800 percent. Our technical community ushered in a new golden age of IBM research and development and earned more United States patent awards than any company for nine years running. We even connected supercomputing with pop culture when a machine named Deep Blue defeated chess grandmaster Garry Kasparov.

In short, once we got back on our feet, shook off the stigma of squandering a seemingly unassailable leadership position, and decided that just maybe our best days were yet to come, the IBM team responded magnificently—just as it had through even the darkest days early in the transformation.[1]

What follows in the next part of this book—the Strategy section—is a change of pace. I could not provide (nor would you want to read) an event-by-event or month-by-month recounting of all that was done to effect the strategic redirection of IBM. My intent is to provide a summary description of the most important strategic changes. Some can be declared successes; others remain works in progress. In each case, what I've included are the moves that were either such distinct departures from IBM's prior direction that they can be considered "bet the company" changes; or those that were so diametrically opposed to the existing culture that they were at great risk of being brought down by internal resistance.

Also, I'll point out that I leave my successor much unfinished business. A number of our strategies are not yet fully deployed; others remain to be defined. More important, the cultural transformation of IBM's formerly successful and deeply entrenched culture—our single most critical and difficult task—will require constant reinforcement or the company could yet again succumb to the arrogance of success.

---

1   See Appendix B for a statistical summary of IBM's performance for 1992–2001.

PART II

Strategy

[12]

A Brief History of IBM

Before we talk about how the new IBM was built, I think it would be helpful to understand, in broad brush strokes, how IBM became the great company most of us revered up until the early 1990s, and what contributed, at least in my view, to its breathtaking decline.

The company's origins go back to the early twentieth century, when Thomas J. Watson, Sr., combined several small companies to form the International Business Machines Corporation. For the first half of the century, IBM's "business machines" embraced a broad and largely unrelated lineup of commercial products; everything from scales and cheese slicers to clocks and typewriters. Of prime importance was the fact that IBM was a pioneer in computation long before most people talked about computers. The company's early electro-mechanical tabulation and punch-card devices introduced computation to business, academia, and government. For example, IBM scored a huge win when it was selected by the United States federal government to help start up and automate the Social Security System in the 1930s.

Inventing the Mainframe

Like Henry Ford, John D. Rockefeller, and Andrew Carnegie, Thomas Watson was a powerful, patriarchal leader who left an imprint on every aspect of his company. His personal philosophies and values—hard work, decent working conditions, fairness, honesty, respect, impeccable customer service, jobs for life—defined the IBM culture. The paternalism engendered by Watson would come to be both an asset and, long past his lifetime, a challenge for the company. However, there's no question that it made IBM highly appealing to a post-depression labor force yearning for job security and a fair deal.

The history that is much more relevant to IBM's turnaround begins with Tom Watson, Jr., who succeeded his father as CEO in 1956 and who boldly brought IBM—and the world—into the digital computer age.

Much has been written about this period and how Tom "bet the company" on a revolutionary new product line called the System/360—the original name of IBM's wildly successful mainframe family.

To grasp what System/360 did for IBM and its effect on the computing landscape, one needs to look no further than Microsoft, its Windows operating system, and the PC revolution. System/360 was the Windows of its era—an era that IBM led for nearly three decades. In fact, the comparison between the IBM of the 1960s and 1970s and the Microsoft of the 1980s and 1990s is most appropriate. Both companies seized upon major technology shifts and brought to market an entirely new capability for customers. Both established commanding market positions and benefited greatly from that leadership.

In IBM's case, the big technology shift came with the advent of the integrated circuit—what we now know as the semiconductor chip. Of course, IBM did not invent the integrated circuit (not any more than Microsoft invented the personal computer!), but Watson and his colleagues understood its significance. Before the integrated circuit, computers were giant, room-size machines, energy inefficient, highly unreliable, and costly to manufacture. Many of these problems could be solved by high-density integrated circuits. Instead of building computers with scores of specialized components, these functions could be miniaturized and packed onto chips.

The new capability that IBM brought to market was the first family of fully compatible computers and peripheral devices. While this hardly sounds revolutionary today, years ago it was a radical concept. Before System/360, IBM was just one of several companies that made and sold computers.

Each company's computers were based on proprietary technology. They didn't work with any other computers, even from the same company, and each computer system had its own peripheral devices like printers and tape drives. This meant that if customers outgrew a computer or wanted the advantages of some new technology, they had to discard all of their hardware and software investments and start over. In today's parlance, they had to "rip and replace" everything.

System/360 represented an entirely new approach. First of all, it would be built with modern, high-performance integrated circuits. This would make the machines simultaneously more powerful, more reliable, and less costly than anything on the market. It would consist of a family of computers—from very small to very large processors—so that customers could make easy upgrades as their needs grew. Software developed for one processor would run on any System/360 processor. All peripheral devices—printers, tape drives, punch-card readers—would work with any processor in the family. For customers, System/360 would be a godsend. For IBM's competitors, it would be a knockout blow.

Of course, envisioning System/360 was one thing. Making it a reality required the equivalent of a man-on-the-moon program. It cost nearly as much. Tom Watson's memoir noted that the investment required—$5 billion (that's 1960s dollars!)—was larger than what the Manhattan Project cost.

Growing Around the Mainframe

Aside from the risk and sheer size of the undertaking, System/360 forced IBM to launch itself into a whole new set of businesses and to develop entirely new sets of skills and capabilities—all of which, in one form or another, still existed by the time I arrived.

IBM had to get into the semiconductor business. Why? Because there was no semiconductor industry yet. IBM had to invest heavily in research and development to create entirely new technologies required for System/360. It's not accidental that this was one of the most progressive periods for IBM research. During this era, IBM scientists and engineers invented the memory chip, the relational database, computer languages such as FORTRAN, and made huge advances in materials science, chip lithography, and magnetic recording.

How did we end up in 1990 with the world's largest software business? Because there would be no usable System/360 without an operating system, or a database, or a transaction processing system, or software tools and programming languages.

Even the sales force had to change. System/360 required a very knowledgeable, consultative sales force that could help customers transform important business processes like accounting, payroll, and inventory management. Traditional order takers couldn't do this job. The company had to create a product service and maintenance capability and a customer-training and educational arm.

Keep in mind that all of this—hardware, software, sales, services—was dedicated and tied to System/360. Despite the fact that IBM, then and now, was regarded as a complex company with thousands of products, I'd argue that, until the mid-1980s, IBM was a *one-product company*—a mainframe company—with an array of multibillion-dollar businesses attached to that single franchise.

And the franchise was a gold mine. IBM's share of the computing market skyrocketed. Competitors reeled; many disappeared. The company's revenues grew at a compound growth rate of 14 percent from 1965 to 1985. Gross profit margins were amazing—consistently around 60 percent. Market share exceeded an astounding 30 percent, which eventually invited antitrust scrutiny.

How the Culture Evolved

This decades-long run of uninterrupted success ties in with the other closely related, and vitally important, aspect of IBM's recent history. This is about its corporate culture—specifically, the kind of culture that arises in an environment without intense competitive pressure or threats. In IBM's case, I never believed the problem was as simple as complacency or entitlement, though there were elements of both present when I arrived. It wasn't about tens of thousands of people growing soft, risk-averse, and slow, though that's been a convenient way to characterize the IBM of the early 1990s.

The IBM culture was the product of two predominant forces. One we've just discussed in detail—the runaway success of the System/360. When there's little competitive threat, when high profit margins and a commanding market position are assumed, then the economic and market forces that other companies have to live or die by simply don't apply. In that environment, what would you expect to happen? The company and its people lose touch with external realities, because what's happening in the marketplace is essentially irrelevant to the success of the company.

What IBM forgot was that all the trappings of its culture—from behaviors that the company valued and rewarded, to how fast things happened, to the luxury of creating all kinds of pride-inducing em-

ployee benefits and programs—were a function of the franchise created by the System/360. It wasn't really the product of enlightened
management or world-class processes. IBM's dominant position had
created a self-contained, self-sustaining world for the company. IBM
had ridden one horse, and ridden it well. But that horse could carry it
only so far before it broke down.

The other critical factor—one that is sometimes overlooked—is
the impact of the antitrust suit filed against IBM by the United States
Department of Justice on January 31, 1969, the final day of the Lyn-
don B. Johnson administration. The suit was ultimately dropped and
classified "without merit" during Ronald Reagan's presidency, but for
thirteen years IBM lived under the specter of a federally mandated
breakup. One has to imagine that years of that form of scrutiny
changes business behavior in very real ways.

Just consider the effect on vocabulary—an important element of
any culture, including corporate culture. While IBM was subject to the
suit, terms like "market," "marketplace," "market share," "competitor," "competition," "dominate," "lead," "win," and "beat" were systematically excised from written materials and banned at internal
meetings. Imagine the dampening effect on a workforce that can't
even talk about selecting a market or taking share from a competitor.
After a while, it goes beyond what is said to what is thought.

Was the antitrust suit "the" pivotal event that caused the culture
of IBM to break down? No. But did it contribute? Some of my long-
term IBM colleagues believe it did. And if timing is everything, as the
adage says, IBM's was lousy. At virtually the same time that the suit
was finally lifted in the early 1980s (and after years of having the fighting spirit drained from the company gene pool), the industry's "next
big thing" arrived. Whether the company fully understood it at the
time, the downward spiral was about to begin.

The Next Big Thing

That next big thing wasn't the advent of personal computing, which is the popular view. The more imminent threat to the mainframe model started with the rise of UNIX, an "open" operating environment championed by companies like Sun and HP. UNIX offered customers the first viable, economically attractive alternative to IBM's mainframe products and pricing.

In the open, plug-and-play world of UNIX, many, many companies could make parts of an overall solution—shattering IBM's hold on architectural control. Almost overnight IBM was under attack by an army of the so-called "pure play" companies like Sun, HP, SGI, Digital, and all the makers of associated software and peripheral products.

Once you understand that, you begin to comprehend John Akers's big bet on a loosely knit confederation of "Baby Blues." He recognized that the vertically integrated industry was ending, and he believed this shift would ultimately take down his vertically integrated company. He was disaggregating IBM in order to embrace what he thought the new industry model was going to be. As I described earlier, I didn't agree with that path and reversed that direction. But I can understand the thinking behind it.

After UNIX cracked the foundation, the PC makers came along swinging wrecking balls. While it's a gross oversimplification to say that IBM's biggest problems stemmed from the failure to lead in PCs, it's clear that the company failed to understand fully two things about personal computing:

- PCs would eventually be used by businesses and enterprises, not just by hobbyists and students. Because of that, we failed to size up the market properly and did not make it a high corporate priority.
- Because we did not think PCs would ever challenge IBM's core en-

terprise computing franchise, we surrendered control of the PC's highest-value components: the operating system to Microsoft, and the microprocessor to Intel. By the time I arrived at IBM, those two companies had ridden this gift from IBM right to the top of the industry.

[13]

Making the Big Bets

If one were to reduce the story of IBM's transformation over the past decade to the bare essentials, the saga would pivot on two big bets: one on the industry's direction, and one on IBM's own strategy. To understand what we did and why we did it, it's helpful to dial back in time and rejoin the discussion of IBM history where it left off in the previous chapter.

Remember that the 1994 time frame I'm describing falls just prior to the Internet revolution. There was a growing confidence inside IBM that the industry was on the cusp of a fundamental shift—the kind of change to the underlying model of computing that comes along about every ten or fifteen years. When that kind of a shift occurs, the companies that seize the moment and lead the movement do exceptionally well—and everyone else dances to their tune.

In the early 1990s the fortunes of the lead horses, in one way or another, were all related to the PC. Of course, that included the PC makers like Dell and Compaq. But without question the dominant leaders were Microsoft, which controlled the desktop operating system, Windows; and Intel, which made the microprocessors. To illustrate the influence these companies wielded, the tandem of Microsoft's Windows and Intel's chips became known as the "Wintel duopoly."

So there was IBM, the company that had led the prior phase of computing and had invented many of the industry's most important technologies, crawling out of bed every morning to find its relevance marginalized by the darlings of desktop computing. The people who had built the systems used by multinational corporations, universities, and world governments were now following the lead of purveyors of word processors and computer games. The situation was embarrassing and frustrating. However, no matter how miserable the present seemed, the future looked even worse.

Their Real Motive

No one believed the PC companies would be content to be kingpins of the desktop. Their aspirations reached right to the heart of IBM's franchise—the large servers, enterprise software and storage systems that anchored the business computing infrastructure. The very name of the new computing model they envisioned—"client/server" computing—revealed their worldview and bias. The "client" referred not to a person, but to the PC. The "server" described mainframes and other business systems that would be in service of the client—providing applications, processing, and storage support for hundreds of millions of PCs each day.

The PC leaders' pitch to business customers was simple and compelling: "You want your employees to make productive use of your business data, applications, and knowledge, which are tied up on old back-office systems. Right now those systems and your PCs don't work together. Since all of your PCs are already Microsoft and Intel machines, you should put in back-office systems that use the same technology."

It was easy to play out the scenario. The PC leaders would march relentlessly up from the PC into business computing and displace IBM products, along with those of vendors like Sun, HP, Digital Equipment, and Oracle. Many of IBM's traditional competitors threw in the towel and joined the duopoly team. It would have been easy to follow HP and UNISYS and all the rest down this path. All of the pundits who followed the industry saw the dominance of this model as inevitable.

It would also have been easy simply to be stubborn and say that the changeover wasn't going to happen, then fight a rear-guard action based on our historical view of a centralized computing model.

What happened, however, is that we did neither. We saw two forces emerging in the industry that allowed us to chart a very different course. At the time, it was fraught with risk. But perhaps because the other alternatives were so unpalatable, we decided to stake the company's future on a totally different view of the industry.

The first force emanated from the customers. I believed very strongly that customers would grow increasingly impatient with an industry structure that required them to integrate piece parts from many different suppliers. This was an integral part of the client/server model as it emerged in the 1980s. So we made a bet—one that, had we articulated it loudly at the time, would have left our colleagues in the industry rolling in the aisles.

Our bet was this: Over the next decade, customers would increasingly value companies that could provide solutions—solutions that integrated technology from various suppliers and, more important, integrated technology into the processes of an enterprise. We bet that the historical preoccupations with chip speeds, software versions, proprietary systems, and the like would wane, and that over time the information technology industry would be services-led, not technology-led.

The second force we bet on was the emergence of a networked model of computing that would replace the PC-dominated world of 1994.

Let me briefly describe our thinking at the time.

A Services-Led Model

As I stated earlier, I believed that the industry's disaggregation into thousands of niche players would make IT services a huge growth segment of the industry overall. All of the industry growth analyses and projections, from our own staffs and from third-party firms, supported this. For IBM, this clearly suggested that we should grow our services business, which was a promising part of our portfolio, but which was still seen as a second-class citizen next to IBM's hardware business. Services, it was pretty clear, could be a huge revenue growth engine for IBM.

However, the more we thought about the long-term implications of this trend, an even more compelling motivation came into view. If customers were going to look to an integrator to help them envision, design, and build end-to-end solutions, then the companies playing that role would exert tremendous influence over the full range of technology decisions—from architecture and applications to hardware and software choices.

This would be a historic shift in customer buying behavior. For the first time, services companies, not technology firms, would be the tail wagging the dog. Suddenly, a decision that seemed rational and straightforward—pursue a growth opportunity—became a strategic imperative for the entire company. That was our first big bet—to build not just the largest but the most influential services business in the industry.

A Networked Model

The second big bet we placed was that stand-alone computing would give way to networks.

That may not sound like a very big or risky bet today. But, again, this was in the context of the 1994 time frame, well before the Internet became mainstream. The first rumblings of change were there. You could find certain industries, particularly telecommunications, that were buzzing about the "information superhighway," a dazzling future of high-speed broadband connections to the workplace, home, and school. If this kind of "wired world" came about, it would change the way business and society functioned.

It would also change the course of computing in profound ways. For one thing, it was virtually certain that world would be built on open industry standards. There would be no other way to fulfill the promise of ubiquitous connections among all the businesses, users, devices, and systems that would participate in a truly networked world. If that standards-based world came to pass, it would represent a major shift in the prevailing competitive landscape.

In any other industry, we *assume* the existence of common standards. We take it for granted that unleaded gas will work in all gasoline-powered cars. We don't think about plugging in appliances or screwing in lightbulbs or turning on faucets. Everything of this nature works because the various manufacturers and service providers in those industries agreed to common standards long ago.

Believe it or not, that's not how things have worked in the IT industry. Based on my experience, it was the only industry on earth where suppliers built products to be compatible with their own gear but not with anyone else's. Once you bought one part of a manufacturer's product line, you were locked in to everything else they made. Imagine, for example, buying a car and discovering you could purchase new tires, spark plugs, filters, accessories, and even the gasoline only from that car's manufacturer.

Of course, I learned that this proprietary model was rooted in IBM's runaway success of the 1960s and 1970s. Other companies— most notably Microsoft—later emulated and perfected this approach and then doggedly refused to abandon it, for precisely the same reason that IBM initially resisted the tug of the UNIX marketplace. Open computing represented a gigantic competitive threat to any company whose business model depended on its ability to control customers based on "choke points" in the architecture.

Fortunately, by the 1980s there were pockets of radical thought inside IBM that were already agitating for the company to join the open movement. And by the mid-1990s, we'd mounted the massive technical and cultural effort required to repudiate closed computing at IBM and open up our products to interoperate with other industry-leading platforms.

Then along came the networked world. If that interconnected, standards-based world took hold, Microsoft would be the most vulnerable. Its insatiable ambitions notwithstanding, not every piece of digital equipment in the world could be part of one architecture, controlled by one company.

Implications of a Post-PC World

There were further implications of a networked world. The PC would be pushed off center stage. Very fast, high-bandwidth networks would allow many of the PC's functions to be performed by larger systems inside companies and the network itself. This system would allow an untold number of new kinds of devices to attach to networks—intelligent TVs, game consoles, handheld devices, cell phones, even household appliances and cars. The PC would be one—but only one—of many network access devices. And if the world was going to be populated by billions of different kinds of computing devices, there would be huge demand for customized chips to power each of these unique devices.

More important for IBM, increasing numbers of people and enterprises conducting business over networks would drive a corresponding increase in computing workloads. The difficult task of managing all of that free-flowing digital information certainly was not going to be done on desktop computers. Those workloads would have to be handled by large-scale systems—meaning huge demand for computing infrastructure products, in addition to networking gear.

Finally, this new landscape would change who made technology buying decisions. In a PC and client/server world, consumers, end-users, and small-department heads were in the driver's seat. But with the action shifting back to enterprise systems and mainstream business strategies, the decision makers would once again be chief technology officers and senior business leaders—people IBM knew and understood.

All of this wasn't so neat, tidy, and clear to us at the time. But there were indications that the world of computing was indeed shifting in ways that, at least in theory, played to IBM's traditional strengths and assets.

We would have to do an enormous amount of work and take significant risks—from continuing to open up all our products, to building the services business. But even the chance that the game might be thrown open to a new set of leaders was powerfully motivating. We were going to take our fate into our own hands. We were going to play offense.

Services—the Key
to Integration

Some people might have been surprised to read in the previous chapter that IBM placed a big bet on services. I mean, hasn't IBM always been known for its doting customer support? Wasn't superior customer service one of IBM's cornerstone beliefs? Wasn't IBM revered for standing by its customers no matter what happened, at any hour of the day or night?

As a customer, I always valued IBM's attentive service. It made up for certain IBM products that weren't quite as powerful or cost-effective as others on the market, and it (almost) justified IBM's very high prices.

But that is not the kind of services I'm talking about here. The big bet we made was on another kind of services—services that would address customer needs that frankly didn't exist during the 1960s, 1970s, and 1980s, prior to the Big Bang of rampant industry disaggregation. Traditionally, IBM's services were completely tied to products—more specifically, products bearing IBM logos. If an IBM system went down, IBM fixed it. However, if customers had a problem with a product from Digital, Compaq, or Amdahl, or if they wanted help installing some other company's equipment, they had to fend for themselves. Services was an adjunct to the main product business.

Enter Dennie Welsh. As with all things in life, luck plays a big part. I got lucky twice at IBM. The first time was at a meeting in 1993 with Dennie, a long-term IBMer, who was running the services organization. The second time was the arrival of the Internet and the big bet we made on the networked world. Coincidentally, Dennie had a strong hand in that, too (more on that later).

When I arrived at IBM, Dennie was running a wholly owned IBM subsidiary named the Integrated Systems Services Corporation. ISSC was our services and network operations unit in the United States—a promising but minor part of IBM's portfolio. In fact, it wasn't even a stand-alone business in IBM. It was a sub-unit of the sales force.

No one forgets a meeting with Dennie. He's a big man, friendly, quick to laugh, but intense. He was a former army pilot and air defense officer who'd made his IBM career in the unit that built highly technical systems for United States government projects, including the Apollo moon program. He was in the control room at Cape Kennedy for Neil Armstrong's historic *Apollo XI* launch to the moon.

It was our first private meeting, but he didn't waste much time on small talk. He told me that his vision of a services company was not one that did just IBM product maintenance and strung together computer codes for customers. He envisioned a company that would literally take over and act on behalf of the customers in all aspects of information technology—from building systems to defining architectures to actually managing the computers and running them for the customers.

My mind was afire. Not only was he describing something I'd wanted when I was a customer (for example, I had tried unsuccessfully to outsource the running of RJR Nabisco's data centers), but this idea meshed exactly with our strategy of integration. Here was a man who understood what customers were willing to spend money on, and he knew what that meant—not just the business potential for

IBM, but the coming restructuring of the industry around solutions rather than piece parts.

However, Dennie pointed out, this system was not going to be easy to implement inside the IBM culture. To be truly successful, we would have to do things that would shake the place to its roots. For example, the services unit would need to be able to recommend the products of Microsoft, HP, Sun, and all other major IBM competitors if that, in fact, was the best solution for the customer. Of course, we'd have to maintain and service these products as well.

Moreover, Dennie believed the services unit would have to be separated from the regular sales force, because our sales force would never permit an IBM services person inside their account if there was any chance that the services rep would sell anything other than IBM products.

Finally he pointed out that the economics of a services business were very different from those of a product-based business. A major services contract might last six to twelve years. An outsourcing contract for, say, seven years might lose money in the first year. All of this was foreign to the traditional world of product sales and would create problems for our sales compensation system and the financial management system.

I left my session with Dennie both thrilled and depressed (a state of confusion I experienced often in my early days at IBM). I was thrilled that I had discovered a base from which we could build the integration capability our customers so desperately needed—and, in so doing, provide the raison d'être for keeping IBM together. I was depressed to realize that despite the powerful logic—that this services-led model was IBM's unique competitive advantage—the culture of IBM would fight it.

Thus began another major challenge: establishing this powerful new business, and at the same time integrating this unit into IBM so that it was viewed, not as a threat, but as a great new ally of our traditional product units.

I knew this was going to be an exquisitely difficult trick to pull off. My experience in prior jobs told me that intense rivalries between units of a large company were a prevalent behavior pattern. The units that had been the traditional base of a company more often than not resisted (overtly or silently) the emergence of a new sibling—either homegrown or an arrival through acquisition.

Building the Organization

Despite Dennie's view that his unit should be a stand-alone business, not a subset of the sales team, I did not break it out initially. Rather, I spent countless hours working with our teams to develop a sense of mutual dependence between the services and sales people. Services people had to learn that the sales team could get them in the door. The sales people had to realize that services specialists could develop major new avenues of revenue in their accounts.

Still, there were fireworks. Throughout those critical early days, it seemed there was a crisis a week between services and some other IBM unit. Many of our brand executives or sales leaders went ballistic every time the services unit proposed a product solution that incorporated a competitor's product. On more than one occasion I found one of these people in my office, railing against the renegades from services. My answer was always the same: "You need to invest the resources necessary to work with the services team to ensure they understand the competitive advantages of your products. View them as a distribution channel for your products. Your competitors do!"

Meanwhile we started to pull the services units together on a global basis. As I said, Dennie ran just a United States services unit. There were dozens of other such services organizations spread all over the world. They had totally different processes, pricing, offerings, terms, and brand names. I asked Dennie to create a unified organization—still under the wing of the sales force—and introduce outsourcing and network services globally. This was a Herculean task—common problem solving, methodologies, nomenclatures, skill definitions, capturing and disseminating knowledge on a global basis, and hiring and training thousands of new people every year.

By 1996 I was ready to break the services unit out as a separate business. We formed IBM Global Services. The change was still traumatic for some of our managers, but it was eventually accepted as inevitable by most of our colleagues.

Had the effort to build IBM Global Services failed, IBM—or at least my vision of IBM—would have failed with it. In 1992 services was a $7.4 billion business at IBM (excluding maintenance). In 2001 it had risen to a $30 billion business and accounted for roughly half of our workforce. I would guess there are few companies that have ever grown a multibillion-dollar business at this pace.

There were several reasons customers were pouring so much investment into services. First, skilled IT professionals were in such short supply that millions of IT jobs went unfilled. Customers simply couldn't staff up to do what needed to be done. But the main reason came back to what Dennie and I had discussed at that first meeting: customers' overpowering desire for someone to provide integration. At first that was just the integration of technologies. But as the networked computing model took hold, it created whole new dimensions around integration, forcing customers to integrate technologies with core business processes, and then to integrate processes—like pricing, fulfillment, or logistics—with one another.

The Nature of the Bet

When I say that we made a big bet on services, let me describe what we gambled.

I have worked in services companies (McKinsey and American Express) and product companies (RJR Nabisco and IBM). I will state unequivocally that services businesses are much more difficult to manage.

The skills required in managing services processes are very different from those that drive successful product companies. We had no experience building a labor-based business inside an asset-intensive company. We were expert at managing factories and developing technologies. We understood cost of goods and inventory turns and manufacturing. But a human-intensive services business is entirely different. In services you don't make a product and then sell it. You sell a capability. You sell knowledge. You create it at the same time you deliver it. The business model is different. The economics are entirely different.

Think for a moment about just the outsourcing business. What you're telling the customer is: "Transfer your IT assets—products, facilities, plus the staff—onto my books. I'll absorb it all, manage it, guarantee performance levels, and promise that you'll always be on or close to the leading edge of technology. All that, *and* I'll charge you less than it's costing you now."

At the same time, you're telling yourself: "I can do all that and still make a profit."

That's a bet on a couple of things, starting with your willingness to use your balance sheet. You can't get into that kind of business without making the commitment to carry the infrastructure and loss until a contract that could extend over five or ten years becomes profitable. There's no such thing as a toe in the water. When you take this plunge, it's full-body immersion.

It's a bet on your ability to drive economies of scale—to consolidate lots of customer data centers into megaplexes (what the industry calls "server farms") or the ability to do with 750 people what two or three customers once did with 1,000.

We had to bet that we could build the recruitment, training, compensation, and HR processes to bring in 1,000 or more people a month—even though we'd never attempted anything remotely close to that. In fact, in the mid- to late 1990s, when services was consistently growing 20-plus-percent a quarter, we knew we could do even better if we had more people. But we capped our hiring at about these levels simply because we thought we'd overextend our ability to hire and train qualified people.

Finally, we had to learn how to be disciplined—how to negotiate profitable contracts, price our skills, assess risk, and walk away from bad contracts and bad deals.

For all of these reasons, I've said repeatedly that this is the kind of capability you can't simply acquire (though our competitors keep trying). The bet you're really making is on your own commitment to invest both the years and the capital, then build the experience and discipline it takes to succeed.

The Future

As I write this chapter in the spring of 2002, the IBM services business is suffering the same slowdown affecting most of the high-tech sector. I am confident this performance slump is temporary because I have never seen a business with such an astounding capacity for self-renewal. Every time the industry moves in a new direction, the IT services opportunity is reinvented. Even in an economic down-turn, many services—outsourcing is the leading example—hold strong appeal as customers look for ways to reduce expenses.

When IBM made its commitment to services, we were playing a bit of a hunch. Today, when I make the statement that this industry, and our company, will be services-led for the foreseeable future, that's no hunch. Since the financial restructuring of IBM began in 1993, ser-vices generated roughly 80 percent of all the company's revenue growth—more than $20 billion of the $25 billion total through 2001.

I can't finish a chapter on the IBM services business without making one additional comment. Early on, when Dennie Welsh was first building his services unit, he went to my predecessor and told him he had to have IBM's number-one sales leader, a person with stature and charisma, a deal-closer extraordinaire. Sam Palmisano, the man who eventually succeeded me as CEO, was running part of IBM's Asian operations at the time. He got the call and became president of ISSC. Not only did Sam take the business to another level, but he was a strong role model for many executives who needed to understand that a significant part of our future was in services.

[15]

Building the World's Already Biggest Software Business

If we were right about the end of one computing era and the arrival of the next, we needed answers to important questions: Where would the value shift in that new environment? Where would the strategic high ground be? What would dominate customers' attention (and spending) the way the PC had during the prior phase?

Certainly networking gear, to keep oceans of digital content moving at high speed and high bandwidth, would be in high demand. To handle the explosion in transactions, customers would need increased server and storage capacity. To design and implement networked solutions, they would need a range of services.

But the linchpin seemed to be software. I'm not referring to the software of the prior era—desktop operating systems and productivity applications sold in shrink-wrapped boxes. The software that would matter in the future would have a very different set of characteristics.

For starters, it would have to be based on open standards that every competitor could use and build on. Why? Because the networked world would have to connect hundreds of millions, eventually billions, of devices and systems. Customers would never permit one company's technology to control all of those connected elements, even if the technology were capable of it.

I'm not a technologist, so I'm not going to try to explain the Internet's underpinnings. Suffice it to say that the Internet is built on a set of open technical specifications called protocols. Once computer systems adhere to those specs, they can connect and become part of the Net.

This was basically client/server's utopian promise—seamless connectivity. Of course, it didn't work. Now, the Internet offered to fulfill that promise, and on a global scale.

Diamond in the Rough

In 1993 very few people—even knowledgeable business executives—would have correctly answered the following question: "What is the biggest software company in the world?" I suspect nearly all would have answered "Microsoft." In fact, IBM sold more software in 1993 than did anyone else.

Why the misperception? It was due mainly to the fact that IBM never thought of itself as a software company, did not talk about itself as a software company, did not have a software strategy, and did not even have a unified software organization.

Software, to IBM, was simply one part of a hardware-based offering. Since every computer needs an operating system, and most need databases and transaction processing capability, IBM built many of these software assets but never viewed them as a unique business. Rather, they were buried inside IBM hardware or sold as an add-on feature. And critically, none of this software worked with computers made by manufacturers other than IBM.

So, problem number one: We didn't have a software mentality, much less a real software business. Problem number two: Most of what we had was built for the mainframe world at a time when the bulk of the customer investment was being made in smaller, distributed systems. Problem number three: a troubled child named OS/2.

My consumer packaged goods background helps me understand the emotional attachment companies have for their products. But the situation is different, and far more intense, in the IT industry. I didn't fully understand this when I came to IBM, but I learned in a hurry when I was thrust into our own religious war—the fight for desktop superiority, pitting IBM's OS/2 operating system against Microsoft's Windows. It was draining tens of millions of dollars, absorbing huge chunks of senior management's time, and making a mockery of our image. And in the finest IBM fashion, we were going to fight to the bitter end.

IBM had always designed its own operating software to run on its hardware. However, when the PC came along, IBM's lack of real commitment to that market resulted in the company's asking Microsoft to provide the operating system for the first IBM Personal Computer. Microsoft seized that miscalculation and artfully built the most powerful franchise in computing.

The highest levels of IBM executives were almost obsessed with the effort to unwind the decisions of the 1980s and take back control of the operating system from Microsoft (and, to a lesser extent, gain control of the microprocessor from Intel). From my perspective, it was an extraordinary gamble for a company to be taking at a time when it was in such a weak financial state.

The pro-OS/2 argument was based on technical superiority. I can say without bias that many people outside IBM believed OS/2 was the better product. The anti-Windows argument was that the legendary Microsoft hype machine was using clever marketing and wily PR to foist an inferior product on consumers, take greater control of the industry, and, in the process, destroy IBM.

What my colleagues seemed unwilling or unable to accept was that the war was already over and was a resounding defeat—90 percent market share for Windows to OS/2's 5 percent or 6 percent.

Not only were we banging our heads against a very hard, unrelenting wall, but I had to wonder if anyone was paying attention to the strategic direction we were talking about. If we truly believed that the reign of the PC was coming to an end, why were we pouring energy, resources, and our image into yesterday's war? Desktop leadership might have been nice to have, but it was no longer strategically vital. Continuing to chase it was more than an expensive distraction, not to mention a source of considerable tension with customers. It was counter to our view of where the world was headed.

The last gasp was the introduction of a product called OS/2 Warp in 1994, but in my mind the exit strategy was a foregone conclusion. All that remained was to figure out how to withdraw. I asked for alternatives and was presented with three. The first two would have involved fairly abrupt terminations of the product line. The third involved a five- to six-year winding down that would cost us hundreds of millions of dollars but would provide support to allow customers using OS/2 to migrate to Windows- or UNIX-based systems in a more manageable fashion. I think you know the decision a former customer made, and IBM today is providing support for customers who still depend on OS/2.

The OS/2 decision created immense emotional distress in the company. Thousands of IBMers of all stripes—technical, marketing, and strategy—had been engaged in this struggle. They believed in their product and the cause for which they were fighting. The doomsday scenario of IBM's losing its role in the industry because it didn't make PC operating systems proved to be little more than an emotional reaction, but I still get letters from a small number of OS/2 diehards.

A Future After the War

With yesterday's war behind us, it was easier to start planning for what lay ahead. As I took inventory of what we had available to us inside IBM, it was a mixed picture: a software business that was big but fragmented and unmanaged; a software portfolio that was closed in a world destined to be open; software built for mainframes, rather than for smaller and more widely dispersed systems; and a business that, aside from operating systems tied to the hardware, was losing huge sums of money.

We needed far more focus. Toward the end of 1994, I decided to pull together all of IBM's software assets under a single executive and ask him to build a distinct, stand-alone software business. John Thompson had attracted my attention very early as one of the most thoughtful and capable managers at IBM. He demonstrated a deep understanding of the technology, had a bright and thoughtful intellect, and, perhaps most important for me, he was able to translate arcane technology into business terms.

At the time, John was running our Server Group—the heart of the company. We were managing a critical technology transition, and he'd been in his position for only about fourteen months, so he was shocked when I asked him to take up the task of creating a new business from scratch. But as he did many other times during my tenure at IBM, he accepted the challenge and brought his many talents to bear quickly and effectively.

It is really difficult to exaggerate the enormity of the problem that John faced as 1995 began. IBM had 4,000 software products, all of which were branded with separate names (most of which were unmemorable and un-"rememberable"). They were made in more than thirty different laboratories around the world. There was no management system, no model for how a software company should run, and no skills in selling software as a separate product.

Over the next two years John and his colleagues recruited and trained 5,000 software sales specialists; they became the backbone of a new sales function in IBM that eventually reached 10,000 by the year 2000.

John reduced the thirty labs to eight and consolidated sixty brands to six. He built a management team, developed strategies, and created marketing programs. His team redirected hundreds of millions of dollars of research expenditures and, in particular, shifted substantial sums of money into the new marketing and sales functions. IBM salespeople loved to sell hardware, and why not? That was how they made their quotas and their money. They had little appetite or skill to go up against Oracle or Computer Associates salespeople who were trained exclusively to sell software.

What remained was to find a sense of direction, a focus, a leadership position that would send a message that IBM was serious about software. John thought he had the answer.

To set up the software bet we were about to make, think about software doing its work on three levels:

- At the base, there are the operating systems that tell the hardware what to do.
- At the top, there is all the application software, like a spreadsheet, a program for calculating your income tax, or a graphic design program. This is what an end user sees on the screen.
- In between, there is a collection of software products that connect the two.

At the base: Microsoft owned the dominant operating system, which, regardless of the fate of OS/2, we believed would become increasingly commoditized in a world of open standards.

At the top: Companies like SAP, PeopleSoft, and JD Edwards dominated the applications software market, while we were an unimportant player.

In the space between: products like databases, systems management software, and transaction management programs. It was the complex, largely invisible layer (aptly named "middleware") about which only the most hardcore techies could get excited.

Yet the more we considered what was going to matter if client/server computing gave way to networked computing, middleware started to look less like a backwater and more like the key strategic battleground. We couldn't see the entire picture at the time, but we could see enough. More users. More devices. More transactions. And more demand for ways to integrate applications, processes, systems, users, and institutions. No operating system was going to be able to tie it all together. But middleware existed to do exactly that.

To provide this kind of integration, however, middleware was going to have to work on all of the major vendors' computer systems that would be linked together over vast new networks. In the industry's jargon, the new middleware would have to work "cross platform," and this represented a major obstacle that we would have to overcome. Up until this point in 1995, all of IBM's software was proprietary and worked only with IBM hardware and other IBM software.

Thus, we launched a massive, multi-year effort to rewrite all of our critical software, not only to be network-enabled, but to run on Sun, HP, Microsoft, and other platforms. It was a hugely expensive and complicated project and it created many of the same internal tensions that our services strategy had evoked. Once again we were collaborating with the enemy!

Acquiring Lotus

In early 1995 John came to me with a bold idea to accomplish two objectives: fill a hole in our middleware portfolio, and plant a flag firmly in the world of collaborative, rather than stand-alone, computing. His idea: to acquire Lotus Development Corporation. He has told me since then that he did it with great trepidation, because it was very much out of character for IBM to "buy" technology rather than build it. Besides that, John and I really didn't know each other very well, and he was asking me to sign a very big check.

Lotus had made its name with its popular 1-2-3 spreadsheet software. But what we wanted most, the crown jewel, was an elegant product called Notes—pioneering software that supported collaboration between large numbers of computer users.

By May John convinced me that IBM should acquire Lotus. Thus began the largest software acquisition in the history of the industry. John had approached Lotus CEO Jim Manzi several times about a possible deal, but he had been turned down. We decided to launch an unsolicited bid. I called Manzi on June 5 to inform him that we were initiating a hostile takeover. I can't be sure, but I do not believe he was surprised. His reply was noncommittal, cold but polite—and very brief.

Anyone will tell you that software acquisitions are risky. The asset you're acquiring is human. If the critical people decide to walk (and a lot of them would certainly have the financial wherewithal to do that once the deal closed), then you've spent a lot of money for some buildings, office equipment, and access to a customer-installed base.

That part of the deal didn't scare me. At American Express we'd acquired First Data Resources, which had a very distinct culture, was privately owned, and didn't want any part of being assimilated into a great big company. I knew it could be managed. But in the case of a maverick software company, I also had a feeling that the effort to win over the workforce was a battle that had to be won before the first shot was fired. We understood that our every move was going to be scrutinized by the Lotus employees, whose trust we desperately needed to win.

We knew that the Lotus workforce—much more than IBM's at the time—was an Internet-centric culture. We mixed the Internet and the IBM home page for fast, unfiltered delivery of our position direct to Lotus employees and shareholders. One minute after I'd ended the call to Manzi, our message, including the letter we'd faxed to Lotus, was live on the Net. As expected, the Lotus workforce knew where to look. They came and they read. We had crossed the first hurdle in the merger of two diametrically opposed corporate cultures.

Privately we were still afraid the deal could take months to complete. Hostile takeovers often become weighty with drama, complete with white knights, court battles, and golden parachutes. But our offer made good strategic sense for everyone, and I think we took all the right precautions. On Tuesday Manzi called me and we talked over dinner in Manhattan that night. Our two companies met in small groups over the next two days to talk about culture, legal issues, employees, and pricing. In a law office on Friday night, we shook hands over the final price: $3.2 billion. By Sunday the boards on both sides had approved the deal. In one week we had wrapped up the biggest software deal in history.

This was also the first hostile takeover enacted by IBM—as well as something of a novelty in the business world. *The New York Times* said: "Perhaps the most striking aspect of IBM's takeover bid, and the one that says the most about these times, is that it defies the accepted wisdom on the difficulties of trying to acquire a company whose primary value isn't in its machinery or real estate but rather, in that most mercurial of assets, people." Fortunately, we were able to keep all of the key people, including Ray Ozzie, the development genius behind Notes. (Jim Manzi stayed on for a few months, but he wasn't the kind of person who was comfortable in a large, complex enterprise.)

There were approximately 2 million Notes "seats" installed in customer locations when the deal closed, in July 1995. That grew to 90 million at the end of 2001. And while the Internet has subsequently obviated much of the need for basic collaborative software, Lotus remains in the sweet spot of the world of knowledge management and collaboration.

In the end we gained more than a software company. Culturally we proved that we could keep some organizational distance and allow a fast-moving team to thrive. Perhaps most important, the hostile acquisition sent a clear signal inside and outside IBM that we were out of survival-mode status and serious about reclaiming a position of influence in the industry.

About nine months later, and again at the urging of John Thompson, we made another big software acquisition—Tivoli Systems—that leapfrogged us into the market for distributed systems management products (more gorpy, but very critical, middleware). Tivoli was a $50 million company when we bought it. Its revenues, augmented by some business we transferred from IBM, are now in excess of $1 billion.

As I write this, IBM's Software Group is one of the most powerful software companies in the world and is positioned as the leading software company in networked computing. With 2001 revenues of $13 billion (second only to Microsoft) and pre-tax profits of about $3 billion (growing at a double-digit rate), we are number one or number two in every segment of the market in which we participate.

The IBM software story is a wonderful microcosm of the overall turnaround that took place at the company over the past decade. Incredible technical resources were unleashed to deploy across the entire industry. The catalyst to drive this transformation was an external event—the arrival of the Internet. Spurred on by this emerging opportunity, we rapidly restructured our assets and organization, and poured resources into acquisitions and development strategies.

Opening the Company Store

So far the logic underlying our big strategic bets has been fairly straightforward. If you're going to be a company that designs, builds, and delivers integrated technology solutions, you need a services capability. If you already develop and sell more software than any other company does, and if you believe software is going to be the connective tissue of the networked world, you ought to run your software business as, well, a business.

But deciding to sell your leading-edge technologies to your own competitors? Imagine having the following conversations:

You are talking to IBM's top technical leaders. They are renowned not only within IBM but throughout their fields. Many have been inducted into the most prestigious scientific and technical academies and bodies. Most have devoted their entire careers to seeing one or maybe two breakthrough innovations come to market inside an IBM product. Now, explain to them that you plan to sell the fruits of their labors to the very competitors who are trying to kill IBM in the marketplace.

Imagine having the same conversation with a sales force that's clashing daily with the likes of Dell, Sun, HP, or EMC—companies that would likely be the big purchasers of that technology.

You can appreciate the internal controversy surrounding the decision to build a business based on selling our technology components—the so-called merchant market or Original Equipment Manufacturer (OEM) business. It was just as difficult as deciding to provide services for non-IBM equipment and enabling IBM software to work with competitors' hardware. And the controversy wasn't only within IBM.

Now imagine another conversation. This time you are talking to the senior leaders of companies like Dell, HP, and Sun. On the one hand, you're asking them to buy technology from you, i.e., enter into a business relationship that makes you money—money that you can put into a war chest and use to compete against them in the marketplace.

On the other hand, you're asking them to trust you to supply them with vital components that they will need for their own products, against which you compete. You promise them that if the supply of those critical parts should ever become tight, they won't find themselves over a barrel, or in one. It's easy to understand the emotion and the mutual suspicions surrounding this business.

The announcement in April 1994 that we would mount a serious push to sell our technology on the merchant market was equal parts business pragmatism and another roll of the dice that we could succeed in a business that was just as novel for us as IT services was. Selling components is a vastly different business than selling finished systems. The competitors and buyers are different. The economics are different. We had to build an organization from the ground up. But the upside was compelling.

- The IBM Research Division was far more fertile and creative than our ability to commercialize all of its discoveries. We were underutilizing a tremendous asset.

- Dispersing our technology more broadly would drive our ability to influence the definition of the standards and protocols that underlie the industry's future development.
- Selling our technology would recoup some of our substantial R&D expenditures and open up a new income stream.
- In a post-PC world, there would be high demand for components to power all the new digital devices that would be created for network access.

IBM Research

As I said earlier, it has been well known for half a century that IBM has one of the most prolific and important scientific research laboratories in the world. IBM has more Nobel laureates than most countries do, has won every major scientific prize in the world, and has consistently been the foundry from which much of the information technology industry has emerged.

However, the Research Division of the early 1990s was a troubled place. My colleagues there saw the company being broken up into pieces and wondered where a centrally funded research organization would fit in an IBM that was being disaggregated. When they heard I had decided to keep the company together, the collective sigh of relief that emanated from Yorktown Heights, New York (the headquarters of our Research organization) was almost audible.

One of the obvious but puzzling causes of IBM's decline was an inability to bring its scientific discoveries into the marketplace effectively. The relational database, network hardware, network software, UNIX processors, and more—all were invented in IBM's laboratories, but they were exploited far more successfully by companies like Oracle, Sun, Seagate, EMC, and Cisco.

During my first year at IBM I probed frequently and deeply into the question of why this transfer of technology invention into marketplace performance had failed so badly. Was it a lack of interest on the part of IBM researchers to deal with customers and commercial products? It did not take long to realize that the answer was no.

The major breakdown was on the product side, where IBM was consistently reluctant to take new discoveries and new technologies and commercialize them. Why? Because during the 1970s and 1980s that meant cannibalizing existing IBM products, especially the mainframe, or working with other industry suppliers to commercialize new technology.

For example, UNIX was the underpinning of most of the relational database applications in the 1980s. IBM developed relational databases, but ours were not made available to the fastest-growing segment of the market. They remained proprietary to IBM systems.

Getting Started

The easiest step we could take initially was to license technology to third parties. This process did not involve selling actual components or pieces of hardware and software, but it did allow other companies access to our patent portfolio or our process technology. ("Process technology" is, as the name suggests, the technology required for the IT industry's own manufacturing processes—the skills and know-how to build leading-edge semiconductor and storage components.) This effort—licensing, patent royalties, and the sale of intellectual property—was a huge success for us. Income from this source grew from approximately $500 million in 1994 to $1.5 billion in 2001. If our technology team had been a business unto itself, this level of income would have represented one of the largest and most profitable companies in the industry!

However, this was just the first step to opening the company store.

We moved on from simple licensing to actually selling technology components to other companies. Initially, we sold fairly standard products that were broadly available in the marketplace but that, nevertheless, IBM chose to make for itself. Here we were competing with many other technology suppliers, such as Motorola, Toshiba, and Korean semiconductor manufacturers. The principal product we offered in the market was simple memory chips called DRAMs (pronounced D-RAMS).

Selling commodity-like technology components is a feast-or-famine business driven not so much by customer demand as by capacity decisions made by the suppliers. Our DRAM business made a gross profit of $300 million in 1995 (the feast), then proceeded to lose $600 million three years later (the famine).

We were not naïve about the nature of this business; its cyclicality was well documented. It turned out, however, that the downturn of 1998 proved to be the worst in the history of the industry.

Why were we in the DRAM business? Well, we really didn't have any choice. We had to prove to the world that we were serious about selling technology components. Most of the potential customers for our technology were worried (quite appropriately) that they might begin to depend on us and that we would subsequently decide to turn inward again.

Consequently, riding the DRAM wave was the price of admission for us to enter the technology component marketplace. We had essentially exited the DRAM market by 1999, but by then DRAMs had given us an entry point. Now potential customers worried less about our reliability as a supplier or whether we had a serious commitment to this business.

We were ready to tackle the emerging opportunity in the components business: The change in computing that we've been talking about was driving a fundamental shift in the strategic high ground in the chip industry.

As I've discussed, the action was going to be driven by the proliferation of Internet access devices, exploding data and transaction volumes, and the continued build-out of communications infrastructure. All that was driving demand for chips—and, to our great delight, chips that would have fundamentally different characteristics from the lookalike processors that powered lookalike personal computers.

In this new model, value would shift to chips that powered the big, behind-the-scenes processors. At the other end of the spectrum there would be demand for specially designed chips that would go inside millions, if not billions, of access devices and digital appliances. And in between would be chips in the networking and communications gear.

This is the kind of sophisticated development activity that not only allows great technology companies to shine, but it also generates margins that support the underlying investment necessary to lead. Over the next four years, IBM's Technology Group went from nowhere to number one in custom-designed microelectronics. I'm happy to say that PowerPC has experienced its own renaissance here, quietly reemerging as a simpler, cheaper, more efficient processor used in a wide range of custom applications, including game consoles. Just consider that IBM's contracts with Sony and Nintendo in 2001 hold the potential to produce more intelligent devices than the entire PC industry produced in 2000.

As a result—and here's the important point—IBM, for the first time in its history, is now positioned to benefit from the growth of businesses outside the computer industry. This diversification does not take us away from our core skills; we have simply extended them to entirely new markets.

Our Technology Group is still young and developing. We can't yet declare victory in this, the third of our growth strategies. It is a piece of unfinished business that I leave for my successor.

Nevertheless, while the economic benefits of our Technology Group strategy have been uneven, the underpinnings of the strategy are powerful and potentially huge for IBM. First, if one believes in the theory of building great institutions around core competencies and unique strengths, then exploiting IBM's technological treasure trove is an extraordinary opportunity for the company.

Second, the evidence to date is fairly clear: The two companies that have enjoyed the highest market valuation in the IT industry over much of the last decade have been component manufacturers—Intel and Microsoft. Certainly one derives enormous benefit from its virtually monopolistic position. But there is no doubt that a strategy built around providing fundamental building blocks of the computing infrastructure has proven to be extremely successful in this industry.

[17]

Unstacking the Stack and Focusing the Portfolio

Before you proceed with this chapter, let me make one disclaimer about the upcoming diagram. I am not going to launch into a primer on computing topologies. Instead, I want to use this oversimplified picture of the industry's structure to illustrate the flip side of our work to restore IBM's economic viability. This is about the bone-jarringly difficult task of forcing the organization to limit its ambition and focus on markets that made strategic and economic sense.

The diagram is commonly called "the stack," and people in the computer industry love to talk about it. The stack shows most of the major pieces in a typical computing environment. At the base are components that are assembled into finished hardware products; operating systems, middleware, and software applications sit above the hardware; and it's all topped off by a whole range of services. Of course, it isn't anywhere near as simple as this in real life.

By now you know that IBM's strategy when it birthed the System/360 was to design and make every layer of this stack. But thirty years later the industry model had changed in two fundamental ways. First, very small companies were providing pieces inside the stack that IBM had invented and owned for so long, and customers were purchasing and integrating these pieces themselves.

Second, and just as threatening, two more stacks emerged. One was based on the open UNIX operating platform. The second was based on the closed Intel/Microsoft platform. In the mid 1980s, when IBM had more than a 30 percent share of the industry, the company could safely ignore these offshoots. But by the early 1990s, when IBM's share position was below 20 percent and still falling fast, it was long past time for a different strategy.

We had to accept the fact that we simply could not be everything to everybody. Other companies would make a very good living inside the IBM stack. More important, just to stay competitive we were going to have to mount massive technical development efforts. We couldn't afford not to participate in the other stacks, where billions of dollars of opportunity was being created.

I've already described the results of our decision to expand into the UNIX and Wintel markets—reinventing our own hardware platforms at the same time that we built new businesses in software, services, and component technologies. Just as important, we had to get serious about where we were going to stake our long-term claims inside the IBM stack.

The first and bloodiest decision was the determination that we would walk away from the OS/2 v. Windows slugfest and build our software business around middleware. Before the 1990s came to a close, we made another strategic withdrawal from a software market.

The Stack

IT Consulting	
Systems Integration	
Outsourcing	
Training & Education	Services
Financing	
Maintenance	

Web Sites	Personal Productivity	
e-Commerce	Engineering & Design	Applications
Supply Chain	Customer Relationship Management	Software
HR	Business Intelligence	

Systems Management	
Application & Transaction Servers	Middleware
Collaboration & Messaging	Software
Database	

Operating System			
Memory	Networking	Displays	Systems
Processor		Storage	

Leaving Application Software

For most of its modern history, IBM made and sold hundreds of business applications, for customers in industries like manufacturing, financial services, distribution, travel, insurance, and health care. These were important applications for important customers, yet we were accomplishing little more than losing our shirts. Jerry York conducted an audit that showed over the previous twenty years IBM had invested about $20 billion in application development and acquisition, with a negative rate of return of around 70 percent!

This was—and is—a very specialized segment of the industry: everything from small-business payroll packages to software for automotive design or even the sophisticated software used to simulate biological and genetic activities. It has always been dominated by entrepreneurial companies that bring obsessive focus to their specialties—such as sales force automation or financial services. Interestingly, nobody has ever succeeded in building a broad portfolio.

When I questioned why we stayed in this business, I was told that application software was critical to the total solution (which was true enough) and that our problems were of execution, and therefore fixable. So we changed executives, tinkered with the strategy, and studied whether we should just buy a few of the best firms in the field. The first candidate was going to be SAP.

Three years, a lot of activity, and a few billion dollars later, we still weren't solution leaders, and we weren't getting anything close to a decent return on our huge investments.

However, one thing we *were* doing exceptionally well was irritating the heck out of the leading application providers—companies like SAP, PeopleSoft, and JD Edwards. These companies were in a great position to generate a lot of business for us if they were inclined to have their applications running on our hardware and supported by our services. Why? Because customers often bought the application first, then looked to that software provider to tell them which hardware to run it on. As long as these companies saw us as a rival, we were driving them into the arms of competitors like Sun or HP.

One example: The segment of IBM that produced applications for distribution and manufacturing customers set a stretch goal to increase sales by $50 million (from a base of about $100 million). It ran ads and promotions and sales contests, and it hit its target. In the process, it alienated every software company in that segment of the market. Those companies, in turn, stopped recommending our hardware and contributed directly to a $1 billion decline in sales of one of our most popular products.

By 1999 we were finally ready to admit to ourselves that we could never be as single-minded as application providers that were in business to do just one thing—and do it better than anyone else. We exited application development but saved the very few pieces of software that IBM had successfully developed and marketed in the past. Thousands of software engineers were reassigned to other work, laboratories were closed, and investments were written off or sold.

Important as it was to stop deluding ourselves about our proficiency in this part of the stack, just as important was the message that we were prepared to work with the leading application software developers. What we said to them was: "We are going to leave this market to you; we are going to be your partner rather than your competitor; we will work with you to make sure your applications run superbly on our hardware, and we will support your applications with our services group."

And rather than just having lunch with them and saying "Let's be partners," we structured detailed commitments, revenue and share targets, and measurements by which both parties agreed to abide.

The first company we approached was Siebel Systems, which had a leadership customer relationship management software package. Its CEO, Tom Siebel, was understandably enthusiastic about the prospect of having IBM's worldwide sales force and services organization marketing and supporting his product. But based on what he'd observed of IBM's agility (or lack thereof), Siebel told us he doubted we could structure a deal on his timetable. He bet the IBM team a bottle of fine wine that the whole process would break down due to what he called "cultural impedance mismatch" between Siebel and IBM.

Five days later Tom was picking out a fine Chardonnay. The contract was signed and we announced the relationship and the new alliance program. Over the next two years we signed 180 similar partnerships.

In hindsight this looks like a no-brainer, given that it dramatically improved the economics of our business and was entirely consistent with our overarching strategy of being the premier integrator. Software companies that in the early 1990s viewed IBM as a major competitor are now very important partners. The amount of incremental revenue we realized is in the billions, and we achieved significant market share gains in 2000, then again in 2001.

The IBM Network

Some may think that the task of moving data from centralized computers to distributed computers, or from one manufacturing site to another, or from one country to another, would be the natural domain of telecommunications companies that had been providing voice transmission for nearly a century. However, until very recently telephone companies had minimal skills in data transmission, and voice services were based on a totally different technology. Moreover, the industry was nationalistic, monopolistic, and highly regulated. Global telecommunications companies did not emerge until the mid-1990s.

So in the spirit of "If they need it, we will build it," IBM in the 1970s and 1980s created multiple data networks to allow its customers to transfer data around the globe. We filled an important void.

By the early 1990s, however, the telecommunications companies were shifting their focus dramatically. Driven in part by deregulation, as well as by the revenue potential of digital services, all of the world's major telecommunications companies were seeking ways to create a global presence, as well as digital capability. In the parlance of both the IT and telecommunications industries, they were talking about moving up the value chain. United States companies that had provided telephone service to customers only in a certain geographic sector of the country were suddenly investing in Latin

American telephone companies. European telephone companies were joining consortia and building wireless networks in remote parts of the world.

In a period of about twenty-four months, the CEOs of nearly every major telecommunications company in the world traveled to Armonk to talk with me about how their companies and IBM could team up to create digital services. The proposals presented to us ran the full spectrum—from modest joint activities to full-blown mergers. However, affiliating IBM in one way or another with a telephone company made no sense to me. I saw little to be gained from a partnership with a regulated company in a different industry. Besides that, we had enough problems in IBM's base business. I wasn't inclined to take on additional challenges.

What did occur to me was that we had an asset that most of these companies would be seeking to build over the next five years. And if the world was moving in the direction we anticipated—toward a glut of many networks (the Internet wasn't even an important consideration at the time)—then the value of our network would never be higher. So we chose to auction it off to the highest bidder. We thought we'd be doing well to get $3.5 billion. But the frenzy eventually produced a bid of $5 billion from AT&T; that was an extraordinary price for a business that produced a relatively tiny percentage of IBM's profits.

This doesn't mean it wasn't a good transaction for AT&T. It allowed AT&T to leapfrog its competitors. But for IBM it was a strategic coup. We got out of a business whose value was going to deteriorate very quickly, as massive capacity was added around the globe. We avoided the huge capital investment to maintain the network. And we exited from another part of the stack that was not strategically vital.

To say there was heavy resistance inside parts of IBM understates the point. People argued, passionately, that we were shortchanging our future. They simply couldn't see the logic in jettisoning a global data network when we all believed we were on the brink of a networked world. Once again there was the "Do it all to be the best" argument. And once again we opted for focus over breadth.

The PC Dilemma

Perhaps the most difficult part of the business that needed to be overhauled during my tenure at IBM was the PC segment of our portfolio. Over the course of nearly fifteen years, IBM had made little or no money from PCs. We sold tens of billions of dollars' worth of PCs during that time. We'd won awards for technical achievement and ergonomic design (especially in our ThinkPad line of mobile computers). But at the end of the day it had been a relatively unprofitable activity. There were times when we lost money on every PC we sold, and so we were conflicted—if sales were down, was that bad news or good news?

The single most important factor in our overall performance was that Intel and Microsoft controlled the key hardware and software architectures and were able to price accordingly. However, we weren't innocent bystanders as they had achieved those dominant positions. We had entered the business in the 1980s with a lack of enthusiasm for the product, as I've already noted. We had consistently underestimated the size and importance of the PC market. We had never developed a sustained leadership position in distribution, vacillating between company-owned stores at one time, to dealers, to distributors, to telephone sales systems. Finally, we couldn't manufacture PCs in a world-class manner in respect to cost and speed to market.

Despite this unacceptable performance, we were never prepared to get out. There were many reasons for this, some more applicable in the early 1990s than they are today. But suffice it to say that, at that time, the PC represented a lot of revenue and critical customer mindshare. In very real ways, a company's PC drove the company's image in the industry. There were raging internal debates about this, but ultimately we felt we couldn't abandon the PC completely and still be the integrator we needed to be for our customers.

So we adopted a strategy of playing to our strengths, primarily in mobile computing and in the market for systems that connected other PCs and helped them function in an integrated way. We waited too long to do it, but we finally abandoned the more commodity-like segments, ceasing to sell to consumers through retail stores and shifting more of our consumer business to direct channels such as ibm.com and telesales. Later on we turned over the development and manufacturing of most of our PCs to third parties, lowering our exposure to this segment even further. Still, it's a spotty record at best, and I am not terribly proud of it.

There were many other steps taken to withdraw from parts of the stack and focus our portfolio. We exited network hardware. Even though we had invented this business, we simply failed to exploit it over the subsequent fifteen or twenty years. We exited the DRAM business. As mentioned earlier, it is a commodity-based, notoriously cyclical market. Midway through 2002 we agreed to divest our hard-disk-drive business through an agreement with Hitachi. As I write these words, other candidates are under active review. This process of selecting markets and competing on the basis of a distinctive, sustainable competency is essential to the new IBM, and I know it will be an ongoing challenge.

Fallacies and Myths and Lessons

As that work proceeds, it is my hope that the company's new leaders keep sight of some of the higher-level lessons that resulted from these decisions.

With os/2—the fallacy that the best technology always wins.
I can understand why this one, in particular, was hard for IBM to accept. In the early days of the computer industry, systems failed frequently and the winner was usually the one with the best technology. So we came to the OS/2 v. Windows confrontation with a product that was technically superior and a cultural inability to understand why we were getting flogged in the marketplace.

First, the buyers were individual consumers, not senior technology officers. Consumers didn't care much about advanced, but arcane, technical capability. They wanted a PC that was easy to use, with a lot of handy applications. And, as with any consumer product— from automobiles to bubble gum to credit cards or cookies—marketing and merchandising mattered.

Second, Microsoft had all the software developers locked up, so all the best applications ran on Windows. Microsoft's terms and conditions with the PC manufacturers made it impossible for them to do anything but deliver Windows—ready to go, preloaded on every PC they sold. (Even IBM's own PCs came preloaded with OS/2 *and* Windows!) And in the mid-1980s, the Windows marketing and PR machine alone had more people than IBM had working with software partners or distributors. Our wonderful technology was whipped by a product that was merely okay, but supported by a company that truly understood what the customer wanted. For a "solutions" company like IBM, it was a bitter but vital lesson.

In the case of application software—the myth of "account control."
This was a term used by IBM and others to talk about how a company maintained its hold on customers and their wallets. It suggests that once customers buy something from a company, then train their people on that product and get familiar with how to support it, it's very hard for them to move to a competitor.

As a former customer, I was always offended and indignant that information technology companies talked about controlling customers. I had this quaint view that it was the job of a supplier to serve customers, not control them!

What IBM has learned from this part of the "unstacking" is that we can be the premier integrator and, at the same time, partner with many other companies in delivering an integrated solution. In fact, wearing my customer hat, I could argue that the role of IT partner, or integrator, cannot be fulfilled by companies that support only one technology or one stack. Beware, customers, of suppliers who provide only UNIX or Wintel answers to your problems. Beware of totally proprietary vendors who fight new developments like Linux. These vendors still view the world through the window of their proprietary stack.

At IBM we now focus on a different stack: the customer's business processes and how we can bring world-class technology—both our own and that provided by other leading companies—to improve those processes.

In the case of PCs, there are still unanswered questions.

Why did we make the decision to exit the application software, network hardware, DRAMs, or the data transmission businesses, but not PCs? Why did we decide to stay in the hardware end of this business, even as we folded our hand on OS/2? In hindsight, was this the right decision? I think it was at the time, but the decision has been painful and costly for IBM.

If there is one lesson to be extracted from this saga, I think it's about staying true to one's strategic vision. I said in 1993 that the marketplace would drive every important decision we at IBM made. But when it came to the PC business, we weren't paying attention to either our customers or our competitors.

One competitor in the PC industry was proving that customers were perfectly willing to buy direct—over the phone, or later via a

Web site. But we were painfully slow to move away from our existing distribution channels. Why? The incomplete and unsatisfying answer at the time was that we'd always done it that way.

I'm not saying the shift to lower-cost, direct channels wouldn't have involved some pain, and I'm not saying that actually sticking to a strategic vision is as easy as articulating it. The tendency, especially in a hyper-competitive marketplace, is to establish a position, hunker down, and defend it. But if we had focused on the marketplace and done our homework, there's no reason the IBM PC business today would be looking up the leader board at Dell.

Opening up our stack (and our minds) to others had many positive effects on IBM. It cut our losses and improved our integrated offerings to customers. And it freed up resources to invest in the future. Huge sums of money and huge quantities of brainpower have been redeployed from wall-banging futility to exciting new work in areas such as storage systems, self-directing computers, bioinformatics, and nanotechnology.

It is all about *focus*—a subject I will return to later.

[18]

The Emergence of e-business

You'll recall that earlier I said I've gotten lucky twice. I described the first piece of luck as my initial encounter with Dennie Welsh, the executive who shared my vision for transforming IBM into a services-led company. Believe it or not, Dennie also played an integral role in my second bit of luck (more on that in a moment).

Long before my arrival at IBM, one of the most widely discussed technology trends in business centered on what was called "convergence"—the melding of telecommunications, computing, and consumer electronics; or, stated differently, the merger of traditional analog technologies with their emerging digital kin. Depending on one's point of view, this either promised or threatened to transform multiple industries.

The subject was not foreign to me. In February 1983, in a speech at the University of Virginia, I made the following observation:

"Computer and telecommunications technologies give us a reach and flexibility that were beyond imagination just a few years ago. . . . Technology has virtually eliminated distance as an obstacle to doing business. Today, an American Express Cardmember from Dallas can buy a plane ticket in Kuala Lumpur and have the purchase authorized in less than six seconds by our computer system in Phoenix, Arizona."

During my brief tenure on the Board of Directors of AT&T, I had learned a great deal about the allure of convergence. That company had bought a computer company, NCR. Some years before that, IBM had bought a telecommunications equipment company, ROLM. Both had placed bets on convergence. And they weren't alone.

If you were a telephone company back then, you were salivating at the prospect of getting beyond dial tone and voice services, to provide all kinds of higher-value services—data, entertainment, and commerce—to homes and businesses.

If you were in the entertainment or media business, convergence represented the ultimate distribution channel. Not only would you be able to digitize all your content, you'd be able to deliver it to devices from personal computers and smart TVs to cell phones and network-enabled wristwatches!

Consumer electronics companies were dreaming up a panoply of devices that would allow billions of people to access this world of digital information and entertainment.

And the information technology industry was gearing up for an explosion of demand for the hardware and software that would manage, process, and store the world's digital content.

So, as I started to probe the strategic thinking inside IBM in 1993, I wasn't surprised to find a number of people who were very excited about this issue. Which brings me back to straight-shooting Dennie Welsh.

Discovering the Cloud

In August 1992 Dennie had landed IBM's largest single contract ever, the outsourcing of all of Sears' data center operations. As part of that contract IBM and Sears merged their private data networks into a joint-venture company called Advantis. It was classic Dennie. He'd been battling IBM to make our global network part of his services organization, but he was getting nowhere with colleagues who couldn't quite see his vision for making the network a big-time profit center. In one stroke he not only closed an $8 billion contract with Sears, but he gained control of more network capacity.

Eventually the joint venture with Sears was dissolved, we took total ownership, and Advantis became part of what we called the IBM Global Network. In its day, IGN was one of the world's most sophisticated networks. It was also the largest Internet service provider in the world—truly an asset without equal, a moneymaker and our stake-in-the-ground in the networking business.

Years later, and as I described in the previous chapter, we divested ourselves of the Global Network in a $5 billion transaction with AT&T. What I didn't mention was that as early as 1993 we knew we'd eventually sell this business. In one of the first conversations I had with Dennie, we agreed that in the long term we'd never be able to justify the massive capital investments required to compete with the telcos. With their assets and base of capital equipment, they could easily undercut our prices. What we couldn't see back then was the Internet, which would entirely obviate the need for us to own a network.

It had to be in one of these early discussions with Dennie that I was introduced to "the cloud"—a graphic much loved and used on IBM charts showing how networks were going to change computing, communications, and all manner of business and human interaction. The cloud would be shown in the middle. To one side there would be little icons representing people using PCs, cell phones, and other kinds of network-connected devices. On the other side of the cloud were businesses, governments, universities, and institutions also connected to the network. The idea was that the cloud—the network—would enable and support incredible amounts of communications and transactions among people and businesses and institutions.

If the strategists were right, and the cloud really did become the locus of all this interaction, it would cause two revolutions—one in computing and one in business.

It would change computing because it would shift the workloads from PCs and other so-called client devices to larger enterprise systems inside companies and to the cloud—the network—itself. This would reverse the trend that had made the PC the center of innovation and investment—with all the obvious implications for IT companies that had made their fortunes on PC technologies.

Far more important, the massive, global connectivity that the cloud depicted would create a revolution in the interactions among millions of businesses, schools, governments, and consumers. It would change commerce, education, health care, government services, and on and on. It would cause the biggest wave of business transformation since the introduction of digital data processing in the 1960s.

So it was natural that when I decided to put someone in charge of a team that would investigate whether we truly believed that convergence was the future—and if so, what to do about it—Dennie got the call. He was passionate on the subject, by then he "owned" our network, and yet I knew Dennie would make sure the team was objective. The team delivered its answer and a set of detailed recommendations in three months.

A Network Computing Blueprint

Dennie's group believed zealously—from the standpoints of technical feasibility and the business opportunity—that this *was* where the industry was headed. But again, what they presented was not primarily an Internet strategy. That's not surprising, because this was back in the days when few people outside of universities and government labs had heard of the Internet. Fewer still believed that the Internet could be a mass-market communications medium, much less a platform for mainstream business transactions. (One notable exception inside IBM was a marketing executive named John Patrick, who had a unique ability to bring the networked world out of the jargon of technology and into the minds of everyday people. John became our spokesperson, demonstrating to our customers and to IBM employees how real-life activities would benefit from the Net.)

Whether we were thinking of networks in general or the Internet specifically, the single most important outcome of Dennie's task force was a recommendation to commit the resources of the entire company to lead this next wave of computing—to mobilize on every front.

On its face, and given the success we'd enjoyed based on proprietary architectures like the System/360, this could have presented a staggering barrier to success. But in truth this one wasn't as tough for us as it would be for some of our mainstream competitors, because we'd long since made our commitment to open, standards-based computing, including embracing all the important Internet standards and protocols. Still, there was plenty to do, much of it described previously.

In software we were fortunate to recognize that middleware would be the integrating glue of networked applications. We had to step up the Internet enablement of these products and develop some new ones.

We'd have to build a significant new services business around what came to be known as Web hosting. And because the networked world was about helping customers transform their businesses, we had to build capability in consulting and implementation services associated with e-business.

In component technology, what began in the early 1990s as a search for a new revenue stream turned into the foundry of specialized chips that would be in high demand.

We also had to fill a lot of gaps. In the summer of 1996, IBM and Lotus announced the Domino Web Server, the first major adaptation of Notes capability for the networked world. We needed an industrial-strength commerce server. After at least one false start we created what is today known as Websphere.

Finally, getting the entire company on board and moving together would be critical. If I asked one unit, such as Global Services or Software, to take the lead, there would be inevitable resistance from the other units. So we decided to form a new group with a pan-IBM mission, headed by an extraordinary executive named Irving Wladawsky-Berger. His job was to evangelize our network strategy across all of IBM's business units, and to get them to change their R&D and marketing plans to embrace the Net. (In what represented at least a minor leap of faith, we called Irving's group the Internet Division.)

Irving was the ideal executive to take on this task. He had come up through the technical side of IBM—through the IBM Research Division and the high-end computing businesses. So he had rock-solid technical credentials. But he also had the ability—in his endearing Cuban accent—to translate very complex technologies into understandable ideas that got people excited. This was important not only because many people in the business units did not fully understand our strategy, but also because Irving would be very effective in influencing people who did not report to him.

I won't minimize the challenges we faced in reorienting the portfolio around the Net. I can say without hesitation, however, that a far bigger and harder piece of work would center on shaping the debate with customers and the industry on where the networked world was headed.

Shaping the Conversation

The first time I made public mention about any of this was to a group of Wall Street analysts in March 1994, several months after my infamous "vision" remark. This time, however, what I said went pretty much unnoticed.

I articulated a strategic vision for IBM constructed around six "imperatives." One of those imperatives was a commitment to lead in what I called the emerging "network-centric world." It was an admittedly clunky piece of terminology, and I can assure you that the next day's news reports were not ringing with anything resembling IBM's bold foray into the uncharted world of networked computing. So be it. By the fall of 1995 my confidence in our strategy had increased to the point where I decided to make network-centric computing the centerpiece of IBM's strategic vision.

In October 1995, *BusinessWeek* published a cover story headlined "Gerstner's Growth Plan: Yes, the CEO does have a vision. It's called network-centric computing." Two weeks later, on November 13, 1995, I gave my first major, inside-the-industry keynote speech at the huge Comdex trade show in Las Vegas. At the time Comdex was the world's largest PC love fest, and a big part of my message was that something called network-centric computing was about to end the PC's reign at the center of the computing universe.

"I assume that all of you have at least one PC," I said. "Most of you probably have several. Unless you're quietly tapping on your notebook while I'm over here talking, all of those PCs are sitting idle—in your briefcase, back in your hotel room, office, in your car, or your home. Think about all that latent computing power that's wasted, totally unused. But in a truly networked world we can share computational power, combine it, and leverage it. So this world will reshape our notions of computing and, in particular, our notions of the personal computer. For fifteen years, the PC has been a wonderful device for individuals. But, ironically, the personal computer has not been well suited for that most personal aspect of what people do: We communicate. We work together. We interact."

Things started happening fast. Netscape had made its headline-grabbing IPO. Microsoft had its epiphany and announced it was committing itself to the Net. There was a buzz building. On the one hand, this was good for IBM, because more voices were now extolling the virtues of a networked world. On the other hand, as more voices joined the conversation and more of our competitors jumped on the networked world bandwagon, the debates and arguments quickly overshadowed the real import and promise of the Net. Microsoft and Netscape waged a titanic battle over browsers. The telcos and new kinds of service providers were racing to connect people and businesses to the Net. Many companies, both inside and outside the IT industry, scrambled to own, leverage, and acquire "content"—news, entertainment, weather, music—thinking that the connected millions would pay to access all of this online digital information.

None of this was good for IBM. Although we wanted to be seen as a leader of this new era, we didn't have a browser. We were already planning to sell the Global Network and exit the business of providing Internet connectivity. We said vocally and proudly that, unlike some of our competitors who were rushing headlong to launch their own online magazines and commerce sites, we were not going to compete with our customers. We weren't going to become a digital entertainment or media company, and we weren't going to get into online banking or stock trading.

Simply put, we had a very different view of what was really happening—about what the Net would mean for business and societal relationships. Terms like "information superhighway" and "e-commerce" were insufficient to describe what we were talking about. We needed a vocabulary to help the industry, our customers, and even IBM employees understand that what we saw transcended access to digital information and online commerce. It would reshape every important kind of relationship and interaction among businesses and people. Eventually our marketing and Internet teams came forward with the term "e-business."

Frankly, the first time I heard it, it didn't do much for me. It didn't mean anything. I didn't think it was particularly memorable. Still, it had potential, and at least it had business, not technology, as its core idea. But we couldn't just plop it into our ads, speeches, and sales calls. We had to infuse the term with meaning and get others in the industry to use it. And we had to strike a balance. We wanted to be seen as the architects of e-business—the agenda setter for this new era, but we decided not to trademark the term "e-business." We wouldn't make it an exclusive IBM term or idea. It was more important to build an awareness and an understanding around our point of view. Creating that environment would require massive investments, both financial and intellectual.

Our executives first unveiled "e-business" during a Wall Street briefing in November 1996. It didn't get a particularly enthusiastic reception. Many months later our advertising agency, Ogilvy & Mather, developed a memorable TV advertising campaign featuring black-and-white office dramas. They worked because they portrayed the confusion most customers felt about the Internet, and they also explained the Net's real value. The commercials were an immediate hit. This was encouraging.

From there we revamped all of our marketing communications—from our trade-show presence to direct-mail campaigns. Every senior executive made e-business a part of his or her presentations and speeches. We communicated frequently on the subject to our people so that they could understand and then evangelize.

To date IBM has invested more than $5 billion in e-business marketing and communications. That's a lot of money, but the returns paid to our brand and our market positioning are incalculable. I consider the e-business campaign to be one of the finest jobs of brand positioning I've seen in my entire career.

In some ways it might have been just a little too successful. In the process of waking up the world to the idea that the Net was about business, we might have inadvertently contributed to the spectacular rise and fall of the dot-coms.

The Emperor's New Economy

In a way that I found astonishing (though by this time I should have known better), almost as soon as the marketplace accepted the idea that the Net was a place where real work would get done, that straightforward idea morphed into the dot-com mania of the late 1990s.

Somehow the Internet came to be seen as some kind of magic wand. In the right hands it could overturn everything from basic economics to customer behavior. It spawned a new class of born-on-the-Net competitors who were going to destroy existing brands and entire industries overnight. Remember the New Economy? It would replace old-fashioned business metrics like profit and free cash flow with "page views," "eyeballs," and "stickiness."

Based on nothing more than a Web site, companies with no earnings and no prospects of ever operating in the black were awarded market valuations that exceeded many of the world's most respected companies. If you weren't a dot-com, prevailing wisdom said you would be dot-toast.

And so here we were, having first advanced the idea that the Net was a medium for real business, watching this wild but totally unsustainable wave of dot-com hysteria crest, then ultimately collapse during 2000. I'd like to say something appropriately high-minded, like: "The courage of our convictions prevented us from joining this fool's-gold rush." In hindsight, the truth is that I'm still a little puzzled by just how easy it was for us to detach ourselves from the dot-com frenzy.

In a speech to Wall Street analysts in the spring of 1999, I had a little tongue-in-cheek fun with the very serious subject of what was real and what was simply unreal.

"These are interesting companies, and maybe one or two of them will be profitable someday. But I think of them as fireflies before the storm—all stirred up, throwing off sparks. But the storm that's arriving—the real disturbance in the force—is when the thousands and thousands of institutions that exist today seize the power of this global computing and communications infrastructure and use it to transform themselves. That's the real revolution."

What did we learn? What were the real lessons, after all the meteoric ascents and equally rapid flameouts?

For customers I think the overriding lesson was that those who didn't get distracted and were willing to do the hard work had a once-in-a-lifetime opportunity—not just to do things better and faster, but to do things that, in fact, they'd never been able to do before. As I write this, customers continue to make major investments to drive transformation through e-business, and they will continue making those investments for the foreseeable future.

For investors as well as customers, the lesson was: no shortcuts. I think for a lot of people, the "e" in e-business came to stand for "easy." Easy money. Easy success. Easy life. When you strip it down to bare metal, e-business is just business. And real business is serious work.

For IBM the lesson was about rediscovering something we'd lost. We found our voice, our confidence, and our ability once again to drive the industry agenda. Our messaging allowed our customers to see benefits and value that were not being articulated by our competitors. The concept of e-business galvanized our workforce and created a coherent context for our hundreds of products and services. The vast new challenges of networked computing reenergized IBM research and triggered a new golden age of technical achievement for the company. Most important, the investment did what we wanted to do at the outset—reestablish IBM's leadership in the industry.[1]

---

1   For more on IBM's view of e-business, see Appendix B.

[19]

Reflections on Strategy

As I look back on the strategic bets we made and how they've played out over the past nine years, I am struck by a conflicting impression. On one level so much about IBM has changed. On another level very little is different.

If you were to take a snapshot of IBM's array of businesses in 1993 and another in 2002, you would at first see very few changes. Ten years ago we were in servers, software, services, PCs, storage, semiconductors, printers, and financing. We are still in those businesses today. Of course, some of those businesses have grown enormously. Others have been refocused. But we have exited just a few segments of the industry. And we have not made gigantic acquisitions to launch us into entirely different industries.

My point is that all of the assets that the company needed to succeed were in place. But in every case—hardware, technology, software, even services—all of these capabilities were part of a business model that had fallen wildly out of step with marketplace realities. There is no arguing that the System/360 mainframe business model was brilliant and correct when it was conceived some forty years ago. But by the late 1980s it had become fatally outmoded. It had failed to adapt as customers, technology, and competitors changed.

What was needed was straightforward but devilishly difficult and risky to pull off. We had to take our businesses, products, and people out of a self-contained, self-sustaining world and make them thrive in the real world.

On a technical level, as I've described, this required the nontrivial task of moving our entire product line—all of our servers, operating systems, middleware, programming tools, and chip sets—from proprietary to open architectures. That alone could have killed us. Many IT companies that have built their businesses on some proprietary product have tried to leap across that chasm. Few have made it across successfully.

This is more than a technical decision to adopt and support a bunch of industry-standard specifications. For IBM, breaking with our proprietary past meant walking away from all the historic architectural control points. It meant stepping onto a competitive playing field that was open to all comers.

The implications of this kind of leap to a company's economic model can be devastating. In IBM's case it meant the collapse of gross profit margins and the attendant changes we had to engineer to lower our cost structure without compromising our effectiveness.

Yet the hardest part of these decisions was neither the technological nor economic transformations required. It was changing the culture—the mindset and instincts of hundreds of thousands of people who had grown up in an undeniably successful company, but one that had for decades been immune to normal competitive and economic forces. The challenge was making that workforce live, compete, and win in the real world. It was like taking a lion raised for all of its life in captivity and suddenly teaching it to survive in the jungle.

This kind of wrenching cultural change doesn't happen by executive fiat. As I found, I couldn't flip a switch and alter behaviors. It was, by any measure, the hardest part of IBM's transformation, and at times I thought it couldn't be done.

PART III

Culture

[20]

On Corporate Culture

B ack in the early 1990s, when a person saw or heard "IBM," what words and images came to mind? "Big computers," "PC," and "ThinkPads," maybe. But inevitably they would also think "big company," "conservative," "regimented," "reliable," and "dark suits and white shirts."

What's interesting is that these latter descriptions refer not to products or services, but to people and a business culture. IBM may be unique in this regard; the company has been known as much for its culture as for what it made and sold. Even today if you pause and think "IBM," chances are you'll think of attributes (hopefully, very positive ones!) of a kind of enterprise and its people rather than of computers or software.

I have spent more than twenty-five years as a senior executive of three different corporations—and I peeked into many more as a consultant in the years before that. Until I came to IBM, I probably would have told you that culture was just one among several important elements in any organization's makeup and success—along with vision, strategy, marketing, financials, and the like. I might have chronicled the positive and negative cultural attributes of my companies ("positive" and "negative" from the point of view of driving marketplace success). And I could have told you how I went about tapping into—or changing—those attributes.

The descriptions would have been accurate, but in one important respect I would have been wrong.

I came to see, in my time at IBM, that culture isn't just one aspect of the game—it *is* the game. In the end, an organization is nothing more than the collective capacity of its people to create value. Vision, strategy, marketing, financial management—any management system, in fact—can set you on the right path and can carry you for a while. But no enterprise—whether in business, government, education, health care, or *any* area of human endeavor—will succeed over the long haul if those elements aren't part of its DNA.

You've probably found, as I have, that most companies say their cultures are about the same things—outstanding customer service, excellence, teamwork, shareholder value, responsible corporate behavior, and integrity. But, of course, these kinds of values don't necessarily translate into the same kind of behavior in all companies—how people actually go about their work, how they interact with one another, what motivates them. That's because, as with national cultures, most of the really important rules aren't written down anywhere.

Still, you can quickly figure out, sometimes within hours of being in a place, what the culture encourages and discourages, rewards and punishes. Is it a culture that rewards individual achievement or team play? Does it value risk taking or consensus building?

I have a theory about how culture emerges and evolves in large institutions: Successful institutions almost always develop strong cultures that reinforce those elements that make the institution great. They reflect the environment from which they emerged. When that environment shifts, it is very hard for the culture to change. In fact, it becomes an enormous impediment to the institution's ability to adapt.

This is doubly true when a company is the creation of a visionary leader. A company's initial culture is usually determined by its founder's mindset—that person's values, beliefs, preferences, and also idiosyncrasies. It's been said that every institution is nothing but the extended shadow of one person. In IBM's case, that was Thomas J. Watson, Sr.

The Basic Beliefs

The defining ethos of Watson, Sr., was palpable in every aspect of IBM. It became part of the company's DNA—from the paternalism to the stingy stock-option program; from the no-drinking-at-corporate-gatherings policy to the preference that employees be married.

Watson's experience as a self-made man engendered a culture of respect, hard work, and ethical behavior. IBM was the leader in diversity for decades, well before governments even spoke of the need to seek equality in employment, advancement, and compensation. A sense of integrity, of responsibility, flows through the veins of IBM in a way I've never seen in any other company. IBM people are committed—committed to their company, and committed to what their company does.

And then there were the more visible, well-known (and, to modern eyes, almost corny) symbols—from the public rituals that celebrated achievement, to the company songs, to the dress code. IBM virtually invented the notion of the company as an all-encompassing context for its employees' lives. And it envisioned its customers in that enveloping way, too.

Of course, enlightened companies and leaders know that an institution must outlive any one person or any one group of leaders. Watson realized this and he deliberately and systematically institutionalized the values that had made IBM under his tenure a very successful company.

He summarized them in what he termed the Basic Beliefs:

- Excellence in everything we do.
- Superior customer service.
- Respect for the individual.

Institutionalizing these beliefs wasn't just a matter of displaying signs in every office (although they were everywhere). The Beliefs were reflected in the compensation and benefits systems, in the management schools, in employee educational and training programs, in marketing, and in customer support. It was the doctrine of the company—and very few companies have extended a doctrine so pervasively.

For a long time it worked. The more successful an enterprise becomes, the more it wants to codify what makes it great—and that can be a good thing. It creates institutional learning, effective transfer of knowledge, and a clear sense of "how we do things." Inevitably, though, as the world changes, the rules, guidelines, and customs lose their connection to what the enterprise is all about.

A perfect example was the IBM dress code. It was well known throughout business circles that IBM salespeople—or, for that matter, any IBM employee—wore very formal business attire. Tom Watson established this rule when IBM was calling on corporate executives who—guess what—wore dark suits and white shirts! In other words, Watson's eminently sensible direction was: Respect your customer, and dress accordingly.

However, as the years went by, customers changed how they dressed at work, and few of the technical buyers in corporations showed up in white and blue. However, Watson's sensible connection to the customer was forgotten, and the dress code marched on. When I abolished IBM's dress code in 1995, it got an extraordinary amount of attention in the press. Some thought it was an action of great portent. In fact, it was one of the easiest decisions I made—or, rather, didn't make; it wasn't really a "decision." We didn't replace one dress code with another. I simply returned to the wisdom of Mr. Watson and decided: Dress according to the circumstances of your day and recognize who you will be with (customers, government leaders, or just your colleagues in the labs).

This codification, this rigor mortis that sets in around values and behaviors, is a problem unique to—and often devastating for—successful enterprises. I suspect that many successful companies that have fallen on hard times in the past—including IBM, Sears, General Motors, Kodak, Xerox, and many others—saw perhaps quite clearly the changes in their environment. They were probably able to conceptualize and articulate the need for change and perhaps even develop strategies for it. What I think hurt the most was their inability to change highly structured, sophisticated cultures that had been born in a different world.

Take the Basic Beliefs. There is no arguing with these. They should be the standard tenets of any company in any industry, in any country, at any period in history. But what the Beliefs had come to mean—or, at least, the way they were being used—was very different in 1993 than in 1962, when Tom Watson had introduced them.

Consider "superior customer service." The supplier–customer power relationship had become so one-sided during IBM's hegemony that "customer service" came to mean, essentially, "servicing our machines on the customers' premises," instead of paying real attention to their changing businesses—and, where appropriate, challenging customers to expand their thinking (as IBM had famously done during the launch of the System/360). We basically acted as if what customers needed had been settled long ago, and our job was to ship them our next system, whenever it came out. Customer service became largely administrative—like going through the motions in a marriage that has long since lost its passion.

The same thing happened to "excellence in everything we do." The pursuit of excellence over time became an obsession with perfection. It resulted in a stultifying culture and a spider's web of checks, approvals, and validation that slowed decision making to a crawl. When I arrived at IBM, new mainframes were announced every four to five years. Today they are launched, on average, every eighteen months (with excellent quality, I might add). I can understand the joke that was going around IBM in the early 1990s: "Products aren't launched at IBM. They escape."

Perhaps most powerful of all the Beliefs—and most corrupted—was "respect for the individual." I am treading on the most sacred ground here, and I do so gingerly. To this day, "respect for the individual" is the rallying cry for the hardcore faithful—for the True Blue, as they call themselves.

But I have to say that, to an outsider, "respect for the individual" had devolved to mean a couple of things Watson certainly did not have in mind. For one, it helped spawn a culture of entitlement, where "the individual" didn't have to do anything to *earn* respect—he or she expected rich benefits and lifetime employment simply by virtue of having been hired.

Or that's the way it appeared to me at first. Later I came to feel that the real problem was not that employees felt they were entitled. They had just become accustomed to immunity from things like recessions, price wars, and technology changes. And for the most part, they didn't even realize that this self-contained, insulated system also worked *against* them. I was shocked, for instance, to discover the pay disparities—particularly in very important technical and sales professions—of IBM employees when compared to the competition and the industry in general. Our best people weren't getting what they deserved.

"Respect for the individual" also came to mean that an IBMer could do pretty much anything he or she wanted to do, within the broad HR and legal rulebooks, with little or no accountability. If you were a poor performer and we terminated you, we weren't respecting your individuality because we hadn't trained you for whatever it was you were expected to do. If your boss told you to do something and you didn't agree, you could ignore the order.

These were very serious problems. They had become deeply engrained through years of self-reinforcing experience. And, most challenging of all, they were almost inextricably interwoven with all that was good, smart, and creative about the company and its people—all the things it would have been madness to destroy, or even to tamper with. We couldn't throw the baby out with the bathwater.

Stepping Up to the Challenge

Frankly, if I could have chosen not to tackle the IBM culture head-on, I probably wouldn't have. For one thing, my bias coming in was toward strategy, analysis, and measurement. I'd already been successful with those, and like anyone, I was inclined to stick with what had worked for me earlier in my career. Once I found a handful of smart people, I knew we could take a fresh look at the business and make good strategic calls or invest in new businesses or get the cost structure in shape.

In comparison, changing the attitude and behavior of hundreds of thousands of people is very, very hard to accomplish. Business schools don't teach you how to do it. You can't lead the revolution from the splendid isolation of corporate headquarters. You can't simply give a couple of speeches or write a new credo for the company and declare that the new culture has taken hold. You can't mandate it, can't engineer it.

What you *can* do is create the conditions for transformation. You can provide incentives. You can define the marketplace realities and goals. But then you have to trust. In fact, in the end, management doesn't change culture. Management invites the workforce itself to change the culture.

Perhaps the toughest nut of all to crack was getting IBM employees to accept that invitation. Many used hierarchy as a crutch and were reluctant to take personal responsibility for outcomes. Instead of grabbing available resources and authority, they waited for the boss to tell them what to do; they delegated up. In the end, my deepest culture-change goal was to induce IBMers to believe in themselves again—to believe that they had the ability to determine their own fate, and that they *already* knew what they needed to know. It was to shake them out of their depressed stupor, remind them of who they were—you're *IBM*, damn it!—and get them to think and act collaboratively, as hungry, curious self-starters.

In other words, at the same time I was working to get employees to listen to me, to understand where we needed to go, to follow me there, I needed to get them to stop being followers. This wasn't a logical, linear challenge. It was counterintuitive, centered around social cues and emotion rather than reason.

Tough as that was, we had to suck it up and take on the task of changing the culture, given what was at stake. I knew it would take at least five years. (In that I underestimated.) And I knew the leader of the revolution had to be *me*—I had to commit to thousands of hours of personal activity to pull off the change. I would have to be up-front and outspoken about what I was doing. I needed to get my leadership team to join me. We all had to talk openly and directly about culture, behavior, and beliefs—we could not be subtle.

[21]

An Inside-Out World

To someone arriving at IBM from the outside, there was a kind of hothouse quality to the place. It was like an isolated tropical ecosystem that had been cut off from the world for too long. As a result, it had spawned some fairly exotic life-forms that were to be found nowhere else. And because IBM was so deeply inbred and ingrown, so preoccupied with its own rules and conflicts, it had lost its robustness. It had become extremely vulnerable to attack from the outside.

This hermetically sealed quality—an institutional viewpoint that anything important started inside the company—was, I believe, the root cause of many of our problems. To appreciate how widespread the dysfunction was, I need to describe briefly some of its manifestations.

They included a general disinterest in customer needs, accompanied by a preoccupation with internal politics. There was general permission to stop projects dead in their tracks, a bureaucratic infrastructure that defended turf instead of promoting collaboration, and a management class that presided rather than acted. IBM even had a language all its own.

This isn't to ridicule IBM culture. Quite the contrary, as I've indicated, it remains one of the company's unique strengths. But like any living thing, it was susceptible to disease—and the first step to a cure was to identify the symptoms.

The Customer Comes Second

I could make the case that during the 1960s and 1970s IBM's self-absorption was actually productive. In those years customers didn't have a lot of insight into what data processing could do for them. So IBM invented these powerful and mysterious machines, and customers looked to us to explain how the technology could be applied to make their companies more efficient. This worldview was, and still is, common throughout the information technology industry. "What we could make" was the starting point, not "what they need."

But as time progressed, businesspeople began to understand the importance of information technology and how it related to everything they wanted to do. I know, because I was one of them. As business strategy began setting the technology agenda rather than the other way around, more and more investments that customers made in IT were being driven by line-of-business managers, not CIOs. Our industry, and IBM in particular, needed to adjust. We needed to open the window to the outside world. And IBM now had competition. UNIX systems and the PC allowed hundreds of competitors to pick away at IBM's franchise. We could no longer run our business like the Roman Empire, confident in our hegemony, certain that those barbarians massing on the borders were no real threat.

And yet I was shocked to find so little customer and competitive information when I arrived. There was no disciplined marketing intelligence capability. What market share data we had was highly questionable, mostly because IBM defined the market, unsurprisingly, in its own image.

We may not have known much about customers, but there was

one group to which we paid plenty of attention: ourselves. In the IBM culture, the organization, and how one fit into the organization, was considered a very important subject. Kremlinology—whereby you judge who was in and out and up and down according to the lineup of leaders atop Lenin's tomb on May Day—was a fine art. For instance, I noticed early on that in any presentation, regardless of subject, the first chart ("foil") invariably depicted the internal organization, including a box showing where the speaker fit on the chart (quite close to the CEO most of the time).

Organization announcements were big deals. When you got promoted, you had a press release, a written internal announcement on our electronic bulletin board, and your boss had a conference call with all of his or her direct reports to announce the good news, with you sitting beside the boss, presumably beaming. Each evening I'd look through my mail and e-mail and find dozens of seemingly minor and innocuous organization announcements, like this one:

The following changes have been announced in the corporate manufacturing and development organization:

- Continuing to report to Patrick A. Toole, IBM senior vice president, manufacturing and development, is Jean-Pierre Briant, IBM director of manufacturing and logistics. Reporting to Mr. Briant are:
  - Jean-Pierre Briant, (acting) IBM director of manufacturing;
  - Lars G. Ljungdahl, who has been named IBM director of logistics and procurement. He was IBM director of logistics processes.
- Reporting to Mr. Ljungdahl are:
  - Lars G. Ljungdahl, (acting) director of IBM order process. His organization remains unchanged.
  - Lars G. Ljungdahl, (acting) director of corporate procurement.

The remainder of Mr. Ljungdahl's organization is unchanged.

I wanted people to focus on customers and the marketplace, not on internal status. A company fighting for survival doesn't need a published caste system with broad readership. So I ended the practice of having a separate category of "IBM" vice presidents v. plain old vice presidents and "IBM" directors, and I banned all press releases about organization.

We had some very enterprising people, however, particularly in our personal computer company. You could always tell when the PC company was about to announce a reorganization, because executives would call the media ahead of time to leak the news and, in the process, make sure reporters understood how well that executive fared in the reorganization. One time *The Wall Street Journal* called and asked that we tell PC company executives to stop calling to leak reorganizations, because the flood of calls was overfilling reporters' telephone mailboxes.

A Culture of "No"

I think the aspect of IBM's culture that was the most remarkable to me was the ability of any individual, any team, any division to block agreement or action. "Respect for the individual" had devolved into a pervasive institutional support system for nonaction.

You saw it at the individual level. One of the most extraordinary manifestations of this "no" culture was IBM's infamous nonconcur system. IBMers, when they disagreed with a position taken by their colleagues, could announce that they were "nonconcurring."

Think about it: At any level of the organization, even after a cross-unit team had labored mightily to come up with a companywide solution, if some executive felt that solution diminished his or her portion of the company—or ran counter to the executive's view of the world—a nonconcur spanner was thrown into the works. The net effect was unconscionable delay in reaching key decisions; duplicate effort, as units continued to focus on their pet approaches; and bitter personal contention, as hours and hours of good work would be jeopardized or scuttled by lone dissenters. Years later I heard it described as a culture in which no one *would* say yes, but everyone *could* say no.

The situation got even worse, because at least a public nonconsent has to be defended among one's peers. More often than not, the nonconcur was silent. It would appear that a decision had been made, but individual units, used to the nonconsent philosophy, would simply go back to their labs or offices and do whatever in the world they pleased!

Here is an internal memo sent to many IBM people in 1994:

The nonconcur process takes place throughout the year with added emphasis on the Spring (strategic) and the Fall (commitment) plan cycles. This note is the official E/ME/A "Kick-Off" of the nonconcur process for this year.

To insure the success of the process I need to know the name of the person in your Division/Industry who will be your Nonconcur Coordinator for the remainder of this year. I will send the detailed Instructions and Guidelines to your coordinator as soon as you identify him/her, and their VNET ID, to me. It would be a great help if you could do this by COB Friday, May 20, 1994.

It is particularly important to be prepared for the issue cycle this year as we are expecting E/ME/A to issue a large number of Nonconcurs during the Spring cycle that are NLS related. Please note that your Product/Industry Managers should be gearing up for issues to reach me, through your coordinator, for the Spring cycle, by Friday, June 3, 1994. (FYI, the Spring Plan has been published by M&D.) Note that I encourage you to get your issues to me prior to the due date, if possible.

Please note that I will not enter any nonconcur into NCMS (Nonconcur Management System) unless you have the timely approval of Bill _____, Bob _____, or their designee, as applicable.

This also means that you must be prepared to help ___/___ escalate, if necessary, through the management chain to Gerstner, if necessary.

Thanks for your help and consideration.

This validation of game-stopping disagreement also manifested itself at the divisional level. Interdivisional rivalries at times seemed more important, more heated, than the battle with external competitors. Early in my IBM career I was shocked when an IBM hardware division struck a deal with Oracle—a company that is an archrival of the IBM software unit—without even telling our software unit in advance.

Don't get me wrong. I'm all for a pragmatic, opportunistic response to complex market realities. I've discussed earlier the need for "coopetition"—whereby we both cooperate and compete with the same companies. But to accomplish that, it takes a mature awareness of who you are as a company, where your deep interests lie, and where they don't. This wasn't that sort of sophisticated ambidextrousness. IBM product salespeople were legendary for going to customers and denigrating another IBM product that might serve in equal capacity in a customer solution. In fact, IBM divisions would bid against one another, and a customer often got multiple IBM bids.

Research-and-development units would hide projects they were working on, so other parts of the company would not learn of them and try to take advantage of their knowledge. It went on and on in a staggering array of internal competition. Teamwork was not valued, sought, or rewarded.

This unique brew of rigidity and hostility often landed on my own doorstep. I discovered that just because I asked someone to do something, that didn't mean the task got done. When I discovered this days or weeks later, I'd ask why. One executive said, "It seemed like a soft request." Or: "I didn't agree with you."

Ironically, at the same time, some people were handing out orders in my name: "Lou said that you should . . ." or "Lou wants you to . . ." Then they followed up on the order so often, so persistently, and so loudly that the tasks actually got done. Unfortunately, many were things I knew nothing about and sometimes *didn't* want done. It got to the point where I had to hold a special meeting with everyone who had access to me for the specific purpose of banning "Lou said" orders on any subject.

Bureaucracy That Hurts

The word "bureaucracy" has taken on a negative connotation in most institutions today. The truth is that no large enterprise can work without bureaucracy. Bureaucrats, or staff people, provide coordination among disparate line organizations; establish and enforce corporatewide strategies that allow the enterprise to avoid duplication, confusion, and conflict; and provide highly specialized skills that cannot be duplicated because of cost or simply the shortage of available resources.

These functions were all critical to an organization like IBM. Coordination was critical because we had a four-way matrix at IBM: geography, product, customer, and solutions. We also desperately needed corporatewide standards for many aspects of the company— e.g., commonality of products around the world for customers who operate globally; and common HR processes so we could move talent quickly and effectively whenever it was needed. And, given the complexity of a highly technical, global company, we clearly needed specialized resources that served the entire company—e.g., branding specialists and intellectual property lawyers.

The problem at IBM was not the presence of bureaucracy but its size and how it was used.

In IBM's culture of "no"—a multiphased conflict in which units competed with one another, hid things from one another, and wanted to control access to their territory from other IBMers—the foot soldiers were IBM staff people. Instead of facilitating coordination, they manned the barricades and protected the borders.

For example, huge staffs spent countless hours debating and managing transfer pricing terms between IBM units instead of facilitating a seamless transfer of products to customers. Staff units were duplicated at every level of the organization because no managers trusted any cross-unit colleagues to carry out the work. Meetings to decide issues that cut across units were attended by throngs of people, because everyone needed to be present to protect his or her turf.

The net result of all this jockeying for position was a very powerful bureaucracy working at all levels of the company—tens of thousands trying to protect the prerogatives, resources, and profits of their units; and thousands more trying to bestow order and standards on the mob.

IBM Lingo

I'm a strong believer in the power of language. The way an organization speaks to its various audiences says a lot about how it sees itself. Everywhere I've worked I've devoted a good deal of personal attention to the organization's "voice"—to the conversations it maintains with all of its important constituencies, both inside and outside the company. And I have chosen my own words—whether in written, electronic, or face-to-face communications—very carefully.

The truth is, you can learn a great deal about a place simply by listening to how it talks. Ordinary discourse at IBM in the early 1990s spoke volumes about the culture's insularity—volumes that were often pretty funny, in a rueful sort of way.

There was a special vocabulary inside IBM—words and phrases used only by IBMers. Also, like the federal government and other bureaucracies, we just *loved* creating and using acronyms, like MDQ, FSD, GPD, and SAA.

As a result, in the early days I would sit through meetings and frequently have no idea what a presenter was talking about. I didn't pretend that I did, however. I'd stop the speaker and ask for a plain-English translation. It was jarring, but people quickly got the point.

Here are some of the most frequently used and colorful IBMisms I heard:

CRISP UP, TWEAK, AND SWIZZLE—Things one had to do to improve a foil presentation

BOIL THE OCEAN—To use all means and options available to get something done

DOWN-LEVEL—Describes a document that has since been improved (or tweaked and swizzled); used most frequently with me when I complained about something: "Lou, you're working from a down-level version."

LEVEL-SET—What you do at the beginning of a meeting to get everyone working from the same facts

TAKE IT OFF-LINE—What two or more people do with an issue that bogs down a meeting—namely, discuss it after the meeting

HARD STOP—A time at which a meeting must end no matter what (I grew to like this expression and use it to this day.)

ONE-PERFORMER—An employee with the company's highest performance rating

MANAGEMENT-INITIATED SEPARATION—Getting fired; commonly used in acronym form: "I've been MISed."

LEFT THE BUSINESS—What an employee who was fired has done

MEASURED MILE—Where a manager puts an employee who, within a year, will most likely leave the business

PUSHBACK—What you run into when someone doesn't agree with you

NONCONCUR—What people do just before pushback

LOBS—Lines of business, or business units (pronounced like what you might do with a tennis ball)

I have always been an advocate for use of plain language that one's customers easily understand, whether it be for invoices, contracts, or simple correspondence. So I decided to begin the end of widespread in-house terms. In an internal memo in 1993, I wrote: "We are also going to take this opportunity to rename some of our organization units to make the nomenclature more understandable, or more transparent, to our customers (call it customer-friendly). Also, we will no longer use the term 'LOB.' Our product units will now be called 'divisions.' "

Presiders Over Process

Soon after I'd joined the company, I asked one of the most senior executives to provide me with a detailed analysis of a major money-losing business at IBM. I did this not only because I wanted the insight from the analysis but also to test this highly rated executive.

Three days later I asked him how the work was progressing. He said, "I'll check with the team and get back to you." At the end of the week, I got the same response: "I'll check with the team leader and let you know" (he later did). When this little scene played out a third time, I finally said, "Why don't you just give me the name of the person doing the work, and from now on I'll speak directly with him or her."

What I discovered was that senior executives often presided. They organized work, then waited to review it when it was done. You were a worker early in your career, but once you climbed to the top, your role was to *preside* over a *process*. Well, my kind of executives dig into the details, work the problems day to day, and lead by example, not title. They take personal ownership of and responsibility for the end result. They see themselves as drivers rather than as a box high on the organization chart.

When I told this senior executive that I expected him to be a direct and active participant in the problem-solving work I asked him to undertake, he was stunned. That was not how he was trained, nor was that expected corporate behavior at the time. The incident was an eye-opener for me. I had an enormous team of executives. I would need to develop a cadre of leaders.

I do not think IBM senior executives would have described their activities quite the way I did. They were simply acting the roles that the long-established interior culture asked them to perform. This is how things were done. It did not mean they were not bright or committed. It was part of a huge, complicated mosaic that had come to define action and behavior. This same situation also extended to the ever-present and powerful administrative assistants, the nonconcur system, and the role of the "corporate officer." (IBM had a practice of electing senior people to the title of "corporate officers." Once you were so recognized, this title stayed with you for life, like tenure at a university. Your performance post-election was not a factor in your continuation in this role.)

[22]

Leading by Principles

In an organization in which procedures had become untethered from their origins and intent, and where codification had replaced personal responsibility, the first task was to eradicate process itself. I had to send a breath of fresh air through the whole system. So I took a 180-degree turn and insisted there would be few rules, codes, or books of procedures.

We started with a statement of principles. Why principles? Because I believe all high-performance companies are led and managed by principles, not by process. Decisions need to be made by leaders who understand the key drivers of success in the enterprise and then apply those principles to a given situation with practical wisdom, skill, and a sense of relevancy to the current environment.

"But what about the Basic Beliefs?" you may ask. "Couldn't they have been revived and turned into the sorts of principles you're describing?" The answer is, unfortunately, no. The Basic Beliefs had certainly functioned that way in Watson's day, then for many decades after that. But they had morphed from wonderfully sound principles into something virtually unrecognizable. At best, they were now homilies. We needed something more, something prescriptive.

In September 1993 I wrote out eight principles that I thought

ought to be the underpinnings of IBM's new culture and sent them to all IBM employees worldwide in a special mailing. In reading them over now, I am struck by how much of the culture change of the following ten years they describe.

Here are the principles and an abbreviated version of how I described each:

1. **The marketplace is the driving force behind everything we do.**

IBM is too preoccupied with our own notions of what businesses we should be in and how they should work. In fact, the entire industry faces this problem. We are all guilty of producing confusing technology and then making it instantly obsolete. IBM has to focus on serving our customers and, in the process, beating the competition. Success in a company comes foremost from success with the customer, nothing else.

2. **At our core, we are a technology company with an overriding commitment to quality.**

There is a lot of debate about what kind of company we are and should be. No need, because the answer is easy: Technology has always been our greatest strength. We just need to funnel that knowledge into developing products that serve our customers' needs above all else. The benefits will flow into all other areas of the company, including hardware, software, and services.

3. **Our primary measures of success are customer satisfaction and shareholder value.**

This is another way to emphasize that we need to look outside the company. During my first year, many people, especially Wall Street analysts, asked me how they could measure IBM's success going forward—operating margins, revenue growth, something else. The best measure I know is increased shareholder value. And no company is a success, financially or otherwise, without satisfied customers.

4. We operate as an entrepreneurial organization with a mini-
    mum of bureaucracy and a never-ending focus on produc-
    tivity.

This will be hard for us, but the new, warp-speed marketplace demands that we change our ways. The best entrepreneurial companies accept innovation, take prudent risks, and pursue growth, by both expanding old businesses and finding new ones. That's exactly the mindset we need. IBM has to move faster, work more efficiently, and spend wisely.

5. We never lose sight of our strategic vision.

Every business, if it is to succeed, must have a sense of direction and mission, so that no matter who you are and what you are doing, you know how you fit in and that what you are doing is important.

6. We think and act with a sense of urgency.

I like to call this "constructive impatience." We are good at re-search, studies, committees, and debates. But in this industry, at this time, it's often better to be fast than insightful. Not that planning and analysis are wrong—just not at the expense of getting the job done *now*.

7. Outstanding, dedicated people make it all happen, particu-
    larly when they work together as a team.

The best way to put an end to bureaucracy and turf wars is to let everyone know that we cherish—and will reward—teamwork, especially teamwork focused on delivering value to our customers.

8. We are sensitive to the needs of all employees and to the com-
    munities in which we operate.

This isn't just a warm statement. We want our people to have the room and the resources to grow. And we want the communities in which we do business to become better because of our presence.

The eight principles were an important first step—not only in defining the priorities of the new IBM, but in attacking the whole idea of management by process. But that first step would be of little value if we couldn't find a way to instill these principles into the DNA of IBM's people. Obviously, exhortation and analysis wouldn't be enough.

What are the levers of motivation? What can a CEO—or, for that matter, a head of state or university president—do to change the attitudes, behavior, and *thinking* of a population? Of course, different people are motivated by different things. Some by money. Some by advancement. Some by recognition. For some, the most effective motivator is fear—or anger. For others that doesn't work; it's learning, or the opportunity to make an impact, to see their efforts produce concrete results. Most people can be roused by the threat of extinction. And most can be inspired by a compelling vision of the future.

Over the past ten years, I've pulled most of those levers.

Waking Up the Leadership Team

In the spring of 1994 I convened my first senior management meeting at a hotel in Westchester County, New York. We had some 420 people there from around the world, representing every part of the company (and a few reporters in the parking lot waiting—in vain—for news). I had one goal more important than anything else: to motivate this group to focus its talents and efforts outside the company, not on one another.

The centerpiece of my remarks began with two charts: one for customer satisfaction, one for market share. The share picture was startling—a loss of more than half our share since 1985 in an industry that was expanding rapidly. The customer-satisfaction chart was just as depressing. We were eleventh in the industry, trailing some companies that don't even exist anymore! I summarized those two snapshots of our collective performance by saying, "We're getting our butts kicked in the marketplace. People are taking our business away. So I want us to start kicking some butts—namely, of our competitors. This is not a game we're playing. We've got to start getting out in the marketplace and hitting back hard. I can assure you, our competitors are focused maniacally on these charts, and they talk us down constantly."

I showed photos of the CEOs of some of our top competitors. The usual suspects—Gates, McNealy, Ellison, and the like. I then read direct quotes from them belittling IBM, gloating over our fall from grace, and questioning our importance in the industry. For example, this from Larry Ellison: "IBM? We don't even think about those guys anymore. They're not dead, but they're irrelevant."

"What do you think happened to all those points of market share?" I asked. "These guys ripped them away from us. And I don't know about you, but I don't like it. And it makes me angry to hear people say things like that about our company. Every time Visa used to run an ad attacking American Express, I knew what was going to happen the next day. The roof was going to come off the building. The general counsel would send for reinforcements to come into the building to keep people from doing things they shouldn't do. I didn't have to pump up the troops. My job was to keep them from overreacting.

"You know, I have received literally thousands and thousands of e-mail messages since I've been in this company, and I've read every one. I want you to know that I cannot—I *cannot*—remember a single one that talked with passion about a competitor. Many thousands of them talked with passion about other parts of IBM. We've got to generate some collective anger here about what our competitors say about us, about what they're doing to us in the marketplace. This competitive focus has to be visceral, not cerebral. It's got to be in our guts, not our heads. They're coming into our house and taking our children's and our grandchildren's college money. That's what they're doing.

"One hundred and twenty-five thousand IBMers are gone. They lost their jobs. Who did it to them? Was it an act of God? These guys came in and beat us. They took that market share away and caused this pain in this company. It wasn't caused by anybody but people plotting very carefully to rip away our business."

I expressed my frustration and my bewilderment about the recurring failure to execute and the company's apparently endless tolerance of it.

"We don't demand implementation and follow-up. We don't set deadlines. Or when they're missed, we don't raise some questions. But we do create task forces; and then they create task forces. We don't execute, because, again, we don't have the perspective that what counts outside [the company] is more important than what counts inside. Too many IBMers fight change if it's not in their personal interest. There's a very powerful word in the IBM vocabulary. I've never heard it in any other company. The word is 'pushback.' It's as if decisions are meant to be suggestions. Since I've been here, I've discovered people who are fighting decisions that were made years ago, while our market share continues to decline.

"When you have market share like that and a customer satisfaction record like that, there isn't a lot of time for debate. We've got to get out and start winning in the marketplace," I said. "This is going to be a performance-based culture. I am personally involved in filling all the new key jobs in this company, because I'm looking for people who make things happen, not who watch and debate things happening."

I shared my feelings about our opportunities and our prospects. I said I considered the people in the room to be the finest collection of talent assembled in any institution in any industry, and that after one year into the job I was convinced that IBM had virtually unlimited potential—but only if we were willing to make the changes I'd laid out. I then outlined the behavioral changes we needed to make in our culture (see next page).

"There are no ifs in my vernacular," I said. "We are going to do it. We're going to do it together. This is going to be a group of change agents—people who are imbued with the feeling of empowerment and opportunity, for ourselves and all our colleagues. Those of you who are uncomfortable with it, you should think about doing something else. Those of you who are excited about it, I welcome you to the team, because I sure can't do it alone."

REQUIRED BEHAVIORAL CHANGE

FROM	TO
Product Out (I tell you)	Customer In (in the shoes of the customer)
Do It My Way	Do it the Customers' Way (provide real service)
Manage to Morale	Manage to Success
Decisions Based on Anecdotes & Myths	Decisions Based on Facts & Data
Relationship-Driven	Performance-Driven & Measured
Conformity (politically correct)	Diversity of Ideas & Opinions
Attack the People	Attack the Process (ask why *not* who)
Looking Good Is Equal to or More Important Than Doing Good	Accountability (always move the rocks)
United States (Armonk) Dominance	Global Sharing
Rule-Driven	Principle-Driven
Value Me (the silo)	Value Us (the whole)
Analysis Paralysis (100+%)	Make Decisions & Move Forward with Urgency (80%/20%)
Not Invented Here	Learning Organization
Fund Everything	Prioritize

It was an emotional talk for me, and I hoped my audience received it that way. I could tell that for many in attendance it had been a good meeting—certainly for those who wanted to create change. For the others? Well, everyone at least expressed agreement. But transforming stated intentions into actual results was another matter.

In fact, in the following weeks and months I heard that while most of the executives were very supportive, some had simply been shocked. It wasn't so much my ideas and messages that startled them. It was my delivery—my passion, my anger, my directness ("kicking butt," for instance, and "ripping away our business"). Very un-IBM. Very un-CEO-like.

I wasn't surprised—or sorry. I had made the conscious decision to jolt the audience. I hadn't done anything simply for dramatic effect. Yes, IBM needed a dose of shock therapy and a gut check. But, more immediately, I needed my executive team to understand who I was and what I was like—and I knew only a handful would ever have a chance to work with me face-to-face. For all kinds of reasons that would unveil themselves in the future, I had to let them see my competitive side.

So I did. Anyone who knows me would tell you that this wasn't an act. I like kicking competitors' butts. And I hate, hate, hate losing.

A New Path for Leaders

Soon after the meeting things started to change. I could sense a little excitement and hope. Some executives were beginning to exhibit the sort of personal leadership and commitment to change that I sought.

I needed, though, to provide support and encouragement for these risk takers. They were still surrounded by a lot of Bolsheviks who longed for the old system. The risk takers needed both a symbol and a structure to validate their behavior.

This was the inception of the Senior Leadership Group (SLG). Formed in February 1995, its primary purpose was to focus attention on the topics of leadership and change. We met for several days once every year to discuss company strategy, but we spent an equal amount of time on leadership.

Given the group's symbolic importance—and the need to infuse it continually with fresh thinking—I decided it was crucial that membership not be automatic, not based on title or rank. I wanted living, breathing role models—regardless of their place on the organizational chart or the number of people underneath them. A great software designer could be a leader, or a great marketer, or a great product developer, just as well as a senior vice president.

Size mattered. The 35 or so executives with whom I met regularly was too few—but the 420 who had attended that first meeting was too many. I eventually settled on a cap of 300. No one would have tenure. Every year a triage would take place: My top executive team would meet and reconstitute the group. Individuals would be proposed for membership and would have to garner the support of the entire executive team. The presence of a new SLG member automatically meant that an existing member had recently retired, or someone would be told he or she was no longer performing in a manner commensurate with our expectations for SLG members. Believe it or not, many of the latter, while very disappointed, stayed on, with our encouragement.

There was high turnover, and that was constructive. Of the original members of the Senior Leadership Group, there were only seventy-one remaining at the meeting in March 2002. This lack of stasis at the top—combined with some early, visible departures of executives who could not or would not operate as team players—was important in driving home the imperative of change. Nothing can stop a cultural transformation quicker than a CEO who permits a high-level executive—even a very successful one—to disregard the new behavior model.

I made it a high priority to promote and reward executives who embraced the new culture. It sent a message to all the up-and-coming managers that the path to success now wound through a different landscape.

Specifically, people wanted to know how they could make it into the SLG one day. Our answer was to create a set of common attributes that we wanted all of our leaders to have, and to formalize them as "IBM Leadership Competencies." Just as we had sought to shift from process-centric management to an approach based on general principles—permitting individuals to apply those principles in their own way, as circumstances dictated—similarly, our leadership competencies described some essential qualities but allowed for a rich, diverse leadership cadre of styles, personalities, and approaches.

The competencies (see following page) became the basis for evaluating every executive in the company. It did not take long for people to realize that this was going to be how you got ahead in the new IBM.

Moreover, all executives, including those reporting directly to me, had to "go to school" for three days to work with trained counselors to understand how they were viewed by their colleagues regarding the competencies and to develop personalized programs to improve their skills.

Making It Happen

Although I actively promulgated the principles and built our management training and evaluation around the Leadership Competencies, the new ways of doing things were much less codified than what they had replaced. That was how I wanted it to be—and it did produce a marked change in our leadership's behavior and focus (not to mention some valuable attrition among those who found the new ways unbearable).

IBM LEADERSHIP COMPETENCIES

Focus to Win

- Customer Insight
- Breakthrough Thinking
- Drive to Achieve

Mobilize to Execute

- Team Leadership
- Straight Talk
- Teamwork
- Decisiveness

Sustain Momentum

- Building Organizational Capability
- Coaching
- Personal Dedication

The Core

- Passion for the Business

However, after a couple of years I realized that the cultural transformation was stalling. The problem was not unexpected; it shows up in most institutional revitalizations. More IBMers were buying into the new strategies, and they said they liked the cultural behavior we needed to execute those strategies. But it all remained predominantly an intellectual exercise. People believed in the new IBM, but they were measured and compensated—and continued to work—as if they were still in the old IBM.

I needed to take our new principles and make them come alive for all IBMers. To do that I needed to make them simpler and bake them into what people did every day. And since people don't do what you *ex*pect but what you *in*spect, I needed to create a way to measure results.

The initial formulation of the need for more simplicity came in late 1994, after a conversation with one of my colleagues. "Over the weekend, I counted them up, and there are about two dozen things that you want me to wake up in the morning and focus on," he said to me. "I can't do it. I'm not that good. What do you really want people to do?"

Thinking back to the senior management meeting earlier that year, my answer was quick: "Win, execute, and team." Those three words captured the commitments I had teed up at the meeting—and they summed up the most important criteria I thought all IBMers needed to apply in setting their goals. This would, at its most basic level, define our new culture. This wasn't empty cheerleading. I had very specific meanings in mind for each word:

- **Win:** It was vital that all IBMers understand that business is a competitive activity. There are winners and losers. In the new IBM, there would be no place for anyone who lacked zeal for the contest. Most crucially, the opponent is *out there,* not across the Armonk campus. We needed to make the *marketplace* the driving criterion for all of our actions and all of our behavior.
- **Execute:** This was all about speed and discipline. There would be no more of the obsessive perfectionism that had caused us to miss market opportunities and let others capitalize on our discoveries. No more studying things to death. In the new IBM, successful people would commit to getting things done—fast and effectively.
- **Team:** This was a commitment to acting as one IBM, plain and simple.

"Win, Execute, Team" began as a mantra—spread throughout the company via multiple media—and eventually took the form of a new performance management system. Every year as part of our annual planning, all IBMers made these three "Personal Business Commitments" (PBCs), then listed the actions they were going to take in the upcoming year that would fulfill the commitments. The specifics varied by job, of course, but the broad approach was uniform. And the PBC program had teeth. Performance against those commitments was a key determinant of merit pay and variable pay.

Of course, in the end, "making it happen" came down to personal leadership—not just my own but that of hundreds of IBMers who were delighted to throw off the old rigidities and behavior patterns, and to forge a new cultural model. Many of them seemingly burst to the surface, elated to be released from a system that had been stultifying and political.

One person deserves special mention here. After my first failure at finding a new head of Human Resources, I hired Tom Bouchard, who had been the top HR person at U.S. West, Inc., and, before that, at United Technologies Corporation. "Bulldog" comes to mind when you think about Tom: a bright, savvy, practical, and hard-working businessman. He was not a traditional HR type; rather, he was a no-nonsense businessman. More than anyone else, he drove our cultural transformation and therefore deserves recognition as one of the heroes of IBM's transformation.

Declaring Our Moon Shot

Have you ever noticed how the past keeps getting better the further into the future you go? Someone once said that the only paradises we have are those that are lost. I think that person must have worked for a legendary business empire like IBM.

The company's golden age—much of it reality, but at least part of it illusion—had such a powerful hold on the imaginations and the hearts of some IBMers that every change was perceived as a change for the worse. They wanted time to stop, despite the realities of the marketplace and societal change.

Our greatest ally in shaking loose the past, as it turned out, was IBM's own precipitous collapse. However, I knew the memory of that wouldn't last forever. Therefore, rather than go with the usual corporate impulse to put on a happy face, spin things optimistically, and declare the turnaround over as soon as possible, I decided to keep the crisis front and center—not irresponsibly; I didn't shout fire in a crowded company. But I didn't want to lose a sense of urgency prematurely.

There came a time, however, when it was clear to all that the company's life-or-death crisis was over. The prospect of institutional death had helped IBMers break from the past. What would be the means by which we could embrace the future? The answer to that came in our e-business strategy. I have already described it as an integrating program for the company on a strategic and operational level, and it was all that. But the appeal of e-business to me was actually even greater for what it would do internally—for our people.

I decided to declare e-business as our "moon shot," our galvanizing mission, an equivalent of the System/360 for a new era. We infused it into everything—not just our advertising, product planning, research agendas, and customer meetings, but throughout our communications and operations—from my e-mails, broadcasts, and town hall visits to the way in which we measured our internal transformation. It provided a powerful context for all of our businesses. It gave us both a marketplace-based mission and a new ground for our own behaviors and operating practices—in other words, culture.

Most important, it was outward-facing. We were no longer focused on turning ourselves around. We were focused on setting the industry agenda again. We shifted the internal discussion from "What do we want to be?" to "What do we want to do?"

Restless Self-Renewal

I came to realize soon after arriving at IBM that there were—and are—tremendous strengths in the company's culture—characteristics no one would want to lose. If we could excise the bad stuff and reanimate the good, what resulted would be an unbeatable competitive advantage.

As I write this, the battle is not over. IBM has, in effect, undergone vast culture change. The "new Blue"—tied to our e-business strategy and focused on the market's most promising growth opportunities—is beginning to take off. IBMers are energized, motivated, and stimulated as they haven't been in a long time. IBM the Leader—though very different from IBM the Leader of an earlier era—is becoming embedded in the minds of more than 300,000 of the brightest people on the planet.

Where do we go from here? One of two things will happen over the next five years:

- Perhaps we will fall once again into the trap of codification. Win, Execute, Team will become platitudes, the same fate that befell the Basic Beliefs. The SLG will become the IBM Management Committee of yore.
- Maybe, on the other hand, we'll figure out a way to hold on to our new-found edge and agility. Maybe we can practice continual, restless self-renewal as a permanent feature of our corporate culture.

This is something that only a handful of institutions ever achieved over an extended period of time. IBM has forged ahead, through a combination of circumstances, heritage, hard work—and luck—to a position where it is now pioneering a new kind of business enterprise—the counterintuitive corporation. I noted some of its characteristics in my final Letter to Shareholders in our 2001 Annual Report:

> ... big but fast; entrepreneurial and disciplined; at once scientific and market-driven; able to create intellectual capital on a worldwide scale, and to deliver it to a customer of one. This new breed continually learns, changes, and renews itself. It is tough and focused—but open to new ideas. It abhors bureaucracy, dissembling, and politicking. It rewards results. Above all, it covets talent and passion for everything it does.

Building on decades of experience, knowledge, maturity, and character, IBM over the past ten years has begun to develop the ability to handle a very high level of internal complexity and even apparent contradiction. Rather than hiding from conflict or suppressing it, we're learning how to manage it, even benefit from it. This equilibrium can be achieved only when an enterprise has a very sure sense of self.

Sustaining that balance will be tough, but I am optimistic. Something has stirred inside this once sleeping giant. Its people have been reawakened to who they are, what they are, what they can do. Their pride has been reinstilled and their hope regenerated.

Besides, the marketplace we're now living in—the most dynamic, competitive, global economy (not to mention political, cultural, and social environment) in recorded history—will help. As long as IBMers remain focused outward, the world will keep them on their toes.

PART IV

Lessons Learned

What did I learn from my IBM experience? What lessons have I accumulated during the course of more than three decades in business? These are questions I get asked a lot these days. And I always preface my answer with the same concerns and hesitancy: I've never been certain that I can abstract from my experiences a handful of lessons that others can apply to their own situations.

Beyond that highly pragmatic consideration, I have more than a sneaking suspicion that what I am prepared to offer here will not surprise, astonish, or delight the reader who has come in search of something resembling a secret formula, or a directory of timeless revelations.

The work-a-day world of business isn't about fads or miracles. There are fundamentals that characterize successful enterprises and successful executives.

- They are focused.
- They are superb at execution.
- They abound with personal leadership.

If not immutable, these three are at least consistent, through the ups and downs of economic cycles, through changes in the leadership of any particular institution, and through technical revolutions, the likes of which we just experienced with the Internet. They apply to enterprises of all sizes and types: large and small companies, publicly traded and non-profit organizations, universities and, in part, governments. At the end of this part of the book, I will address a final issue that is unique to the largest and most complex institutions: how to strike the appropriate balance between integration and decentralization.

Focus–You Have to Know (and Love) Your Business

F ew people and few institutions would admit to a lack of focus, even in an exercise of honest self-evaluation. However, I have learned that lack of focus is the most common cause of corporate mediocrity. It shows up in many forms, most notably in the items that follow.

"The grass is greener."

This is the most pernicious example. In my thirty-five-year business career I have seen many companies, when the going gets tough in their base business, decide to try their luck in new industries. It's a long list: Xerox going into financial services; Coca-Cola into movies; Kodak into pharmaceuticals.

I remember when I was a student at Harvard Business School forty years ago, a marketing professor argued that the problem with buggy-whip companies was they thought they were in the buggy-whip business, not the transportation business. The professor argued that companies often focus on too narrow a segment and fail to see important changes in their marketplaces. I cannot argue with the underlying premise here, but I *would* argue that it's very, very hard for a buggy-whip company to become an airplane manufacturer.

Too many executives don't want to fight the tough battles of resurrecting, resuscitating, and strengthening their base business—or they simply give up on their base business too soon. As IBM made its bet on the kind of convergence I described in Chapter 18, it diverted its attention from what it always did well—building large-scale, powerful computers—and bought a telephone switching company (ROLM). When things got tough in the charge card and travel business in the 1980s, the chairman of American Express tried to get into the cable TV, entertainment, and book publishing businesses. Of course, American Express brought no skills to any of those businesses. Nabisco, one of the great food companies in the world, bought a tobacco company in 1985. Fourteen years later it spun off the tobacco business, and the only true long-term result was a weakening of the food company.

That's usually what happens when a company strays from its core competencies. Its competitors rejoice at, and prosper from, the lack of focus. And the company ultimately sinks into a deeper hole.

The fact is, in most cases a company has a set of competitive advantages in its base business. It may be hard—very hard—to redirect or reenergize an existing enterprise. Believe me, it's a lot easier than throwing that enterprise over the fence into a totally new environment and succeeding. Age-old common sense: Stick to your knitting; dance with the partner who brought you. History shows that truly great and successful companies go through constant and sometimes difficult self-renewal of the base business. They don't jump into new pools where they have no sense of the depth or temperature of the water.

"We need to grow, so let's go acquire somebody."

Related very much to the ongoing commitment to building a core business is the ability to say no to acquisition fever. This is a contagious disease that infects too many executives. When given a choice of working hard to fix a base business or, instead, completing a glamorous acquisition and crowing about its promise on the financial TV stations, too many executives opt for the latter. As I look back on my IBM life, there is no question that a good portion of our success was due to all of the deals we *didn't* do. A partial list of the companies that were proposed as acquisition candidates includes: MCI, Nortel, Compaq, SGI, and Novell and Telecoms galore. Investment bankers with thick, blue books were always ready to describe a yellow brick road leading to the wonderful city of Oz. Not one of these deals would have worked.

I could tell a lot of investment-banker stories, but perhaps the one that stands out in my mind the most was the proposal from one bank that IBM acquire Compaq Computer. The summary of the transaction that was included in the front of the ever-present blue book showed IBM's stock price going up forever after completing the transaction. Surprised at how this tree would grow to heaven, I rummaged through the appendix and found that IBM's profits for the next five years (roughly $50 billion after taxes) would be wiped out by this transaction and we would show huge losses over that entire period. When I told my CFO to question the banker about how this could be viewed as positive by the investment community, the answer came back: "Oh, investors would all see right through this. It wouldn't matter." Ah, if only the elixir peddled by investment bankers worked, then CEOs would never have to worry or even work. Hello, golf!

Back in the real world, however, there have been numerous, empirical studies conducted over the past twenty years showing that the likelihood of an acquisition's proving to be a failure far outweighs the chances of success. That doesn't mean I think acquisitions have no place in good corporate strategy. IBM made ninety acquisitions during my tenure as CEO. The most successful were those that fit neatly into an organic growth plan. IBM's purchase of Informix is a great example. We were neck-and-neck with Oracle in the database business, and Informix, another database company, had lost its momentum and market leadership. We didn't need to buy Informix to get into the database business or to shore up a weak position. However, we did acquire a set of customers more quickly and more efficiently than we could have following a go-it-alone strategy.

The same was true with a number of other acquisitions in which we basically bought technology that we were going to have to develop ourselves but were able to accelerate our control of that technology through a highly focused acquisition. In other words, acquisitions that fit within an existing strategy have the most likely probability of success. Those that represent attempts to buy new positions in new marketplaces or that involve smashing together two very similar companies are fraught with risk.

Steely-Eyed Strategies

Bottom line: At the end of the day a successful, focused enterprise is one that has developed a deep understanding of its customers' needs, its competitive environment, and its economic realities. This comprehensive analysis must then form the basis for specific strategies that are translated into day-to-day execution.

Sounds simple, doesn't it? Yet, in my experience, not enough companies really do the analytical work in a truly objective way (hope usually prevails over reality); even fewer can then translate the analysis into precise action programs that are tracked month by month.

As I mentioned earlier, perhaps my most infamous quote dealt with the subject of vision. During my years at McKinsey, seeing many different companies, I was always amazed at how many executives thought that "vision" was the same as "strategy." It's very easy to develop visions. It's the same thing as Babe Ruth pointing to the fences. How many Babe Ruths do you think have pointed to the fences in the last twenty years? How many do you think hit a home run within the next minute?

Vision statements can create a sense of confidence—a sense of comfort—that is truly dangerous. Vision statements are for the most part aspirational, and they play a role in creating commitment and excitement among an institution's employees. But in and of themselves they are useless in terms of pointing out how the institution is going to turn an aspirational goal into a reality.

Again, good strategies start with massive amounts of quantitative analysis—hard, difficult analysis that is blended with wisdom, insight, and risk taking. When I first arrived at IBM I asked, "What do our customers think about us? Let me see our customer satisfaction data." I got back reports that were amazingly positive. Basically, customers loved us.

It was all statistical; it all seemed thorough and accurate. However, it just didn't make sense, given that we were losing share in almost every one of our product lines. It took me a while, but I finally discovered that the way we measured customer satisfaction was to ask our sales force to pick some of their customers and ask them to complete a survey. IBM does not hire dumb salespeople. They obviously picked their best and happiest customers, and we were getting lots of positive data and absolutely fooling ourselves every day.

Moreover, every part of IBM was doing its own thing: We were conducting 339 different satisfaction surveys. Disparate methodologies made it impossible to get a single view—even if the sample wasn't biased by the sales force.

Today we conduct fourteen comprehensive customer surveys, administered by an independent research firm. Names are sourced from external lists (not the sales force) and we interview almost 100,000 customers and noncustomers every year. Surveys are conducted in thirty languages in fifty-five countries, and they compare our performance against those of all our major competitors. Most important, the data are incorporated into our tactical and strategic plans on a semiweekly basis.

Intelligence Wins Wars

Perhaps the most difficult part of good strategy is hard-nosed competitive analysis. Almost every institution develops a pride in itself; it wants to believe it's the best. And a lot of what we as managers do is encourage that sense of loyalty and pride. However, this family feeling often gets in the way of really deep competitive insight. We want to believe our products are better than our competitors' products. We want to believe customers value us more than they do our competitors.

Product managers want their bosses to believe the manager has created the best products in the industry. But facts are facts, and they've got to be assembled on a continuous, unbiased basis. Products have to be torn down and examined for cost, features, and functionality. Each element of the income statement and balance sheet has to be examined with total objectivity vis-à-vis competitors. What are their distribution costs? How many salespeople do they have? How are their salespeople paid? What do distributors think about them v. us? There are hundreds of questions that need analytical examination and which then must be pulled together in comprehensive, deep competitive assessments.

Often a root cause of inadequate competitive analysis is asking the innkeeper how good the inn is. It's fairly axiomatic that most managers are not going to strategize themselves out of business. Most managers are not going to present to corporate officers an unvarnished, bleak picture of their stewardship. (Perhaps the only time you really get a totally objective analysis of a business from a division president is when he or she first arrives on the scene. Then there's no accountability for prior mistakes—it's the prior incumbent's problem!)

Good Strategy: Long on Detail

The most important value-added function of a corporate management team is to ensure that the strategies developed by the operating units are steeped in tough-minded analysis, and that they are insightful and actionable. All of the critical assumptions—things such as pricing and industry growth rates—need rigorous and tough-minded review.

Why is this all so important to the subject of focus? Because truly great companies lay out strategies that are believable and executable. Companies that leap into new businesses and chase acquisitions willy-nilly are those that really don't have a conviction about their existing strategy. They don't have a clear understanding of the five or six critical things they need to do in their base business to be successful. Those five or six things are the prime elements that the organization should be preoccupied with every day, then measuring, adjusting, and reallocating resources as necessary.

Again, good strategies are long on detail and short on vision. They lay out multi-year plans in great quantitative detail: the market segments the company will pursue, market share numbers that must be achieved, expense levels that must be managed, and resources that must be applied. These plans are then reviewed regularly and become, in a sense, the driving force behind everything the company does.

Consequently, when an acquisition opportunity shows up from your friendly investment banker, it isn't his or her analysis that is examined. Rather, you do a detailed analysis of how the acquisition fits into the strategy. In fact, if a company hears about an attractive acquisition candidate first from an investment banker, it almost always means the company hasn't done a good job on its strategy. A good strategy will always identify critical holes, competitive weaknesses, and the potential to fill them with tactical acquisitions. I have bought many companies during my business career; I can't remember one of them that was a new idea unearthed by an investment banker.

The Hard Part: Allocating Resources

Finally, making sure that resources are applied to the most important elements of the strategy is perhaps the hardest thing for companies to do. Too many companies view strategy and operations as two separate activities. Strategies are completed once a year, reviewed during long meetings, and approved by some higher authority; then everybody goes off and continues to run the business in much the same way that they did before. If, in fact, a strategy does call for a different set of actions, the very difficult task of taking resources away from some other activity in the company and reassigning them to the higher priority is not done well in many businesses.

Let's return to customer satisfaction at IBM. After we developed truly effective, independent measurements to ascertain how our customers viewed us and our competitors, it was clear that one of our biggest problems was with how easy—or not—it was to do business with us. Our customers liked our products, liked our breadth of experience, liked our ability to help them solve problems, but they often found us to be maddeningly difficult to deal with and/or to get answers from quickly.

Addressing this issue was not easy. There was not a single silver bullet we could fire to solve the problem. There was not a single project we could heap a bundle of money on to make sure the mission got done. It involved hundreds of projects, cutting across the entire company from salespeople to lawyers to telephone clerks.

In some companies such a project, mundane in its day-to-day activities but essential in its strategic context, would die of its own weight and its lack of connectiveness to the daily grind in a relatively short period of time. We had to work hard to maintain its vitality, funding, and focus. It worked, but it was a reminder to me of how difficult it is to get large organizations to give meaningful resources and attention to matters that offer little or no benefit to quarterly results, but which are critical to long-term success.

Survival of the Fattest

Here's my last observation on focus: The Darwinian concept of survival of the fittest unfortunately doesn't work in a lot of companies. Instead, too often the rule is "survival of the fattest." Divisions or product lines that are successful today always want to redeploy their cash and other resources into existing products and existing markets. Finding ample resources to fund new growth and new businesses is one of the hardest tasks of a corporate leader.

While we never reached the level of performance I would have liked at IBM, we worked very hard at the process of starving the losers and investing in new big bets. It required a very different process than the one necessary for developing strategy. It required a rigorous portfolio review in which we said to the entire company: Investment dollars belong to Corporate, all of them, not just the discretionary new capital. We try to start with all of our businesses—successful and not so successful—as a zero-sum planning process every few years. This allowed us to kill thousands of research projects, eliminate hundreds of products, sell large businesses, and redeploy resources into promising new ventures. Even then we could not be sure we had effectively redeployed our assets. Those new ventures had to be protected from the normal budgetary cycle because if things get tight, more often than not, profit-center managers would be tempted to starve the future-oriented projects.

This is not the place to explain the many things we did to avoid the problems and to support new businesses, but it is a very important aspect of my overall conviction that focus is a critical element of institutional success. If a management team doesn't believe that it has identified and is seriously funding new growth opportunities, then it is likely to wander off and drink the heady brew of acquisitions and diversification—and ultimately fail.

[24]

Execution–Strategy Goes Only So Far

E xecution—getting the task done, making it happen—is the most unappreciated skill of an effective
business leader. In my years as a consultant, I participated in the development of many strategies for many companies. I will let you in on
a dirty little secret of consulting: It is extremely difficult to develop a
unique strategy for a company; and if the strategy is truly different
from what others in the industry are doing, it is probably highly risky.
The reason for this is that industries are defined and bounded by economic models, explicit customer expectations, and competitive structures that are known to all and impossible to change in a short period
of time.

Thus, it is very hard to develop a unique strategy, and even
harder, should you develop one, to keep it proprietary. Sometimes a
company does have a unique cost advantage or a unique patented position. Brand position can also be a powerful competitive position—a
special advantage that competitors strive to match. However, these
advantages are rarely permanent barriers to others.

At the end of the day, more often than not, every competitor basically fights with the same weapons. In most industries five or six success factors that drive performance can be identified. For example, everyone knows that product selection, brand image, and real estate costs are critical in the retailing industry. It is difficult, if not impossible, to redefine what it takes to be successful in that industry. Dot-com retailers were a good example of a spectacular failure to understand that you cannot suspend the fundamentals of an industry.

So, execution is really the critical part of a successful strategy. Getting it done, getting it done right, getting it done better than the next person is far more important than dreaming up new visions of the future.

All of the great companies in the world out-execute their competitors day in and day out in the marketplace, in their manufacturing plants, in their logistics, in their inventory turns—in just about everything they do. Rarely do great companies have a proprietary position that insulates them from the constant hand-to-hand combat of competition.

People Respect What You Inspect

At McKinsey my colleagues and I were continually frustrated to see one company after another invest thousands of hours and millions of dollars to develop solid, effective statements of strategic direction and then waste all the time and money because the CEO was unwilling to drive change through the organization. At other times, the CEO *thought* change was taking place in the organization but failed to inspect what, in fact, was going on.

Perhaps the greatest mistake I've seen executives make is to confuse expectations with inspection. I have sat through hundreds of meetings in which strategies—good, solid strategies—have been presented and the business leader has agreed: "Yes, this is what we're going to do." I've seen well-written, sometimes brilliant strategy documents promulgated to the organization. I've seen great video, intranet, and face-to-face messages describing with excitement and passion a new and daring direction for an enterprise. But, alas, too often the executive does not understand that people do what you *inspect,* not what you *expect.*

Execution is all about translating strategies into action programs and measuring their results. It's detailed, it's complicated, and it requires a deep understanding of where the institution is today and how far away it is from where it needs to go. Proper execution involves building measurable targets and holding people accountable for them.

But, most of all, it usually requires that the organization do something different, value something more than it has in the past, acquire skills it doesn't have, and move more quickly and effectively in day-to-day relationships with customers, suppliers, and distributors. All of this spells change, and companies don't like to change because individuals don't like to change.

As I've mentioned earlier, IBM knew what was going on in the computer industry in the late 1980s and early 1990s. It had documented numerous strategies to deal with a changing world. One document described the environment as "a sea of speedboats surrounding a floundering super tanker [IBM]." Newspaper accounts in the early 1990s suggested that my predecessor was exhorting and pressing the company to pursue new strategies. So what happened? The strategic requirements were clear, the CEO was demanding their implementation, but the company stood still in the water.

Execution is the tough, difficult, daily grind of making sure the machine moves forward meter by meter, kilometer by kilometer, milestone by milestone. Accountability must be demanded, and when it is not met, changes must be made quickly. Managers must be asked to report on their performance and explain their successes and failures. Most important, no credit can be given for predicting rain—only for building arks.

I believe effective execution is built on three attributes of an institution: world-class processes, strategic clarity, and a high-performance culture. Let me touch briefly on each.

World-Class Processes

Earlier in this section I mentioned that in every industry it is possible to identify the five or six key success factors that drive leadership performance. The best companies in an industry build processes that allow them to outperform their competitors vis-à-vis these success factors. Think about great companies: Wal-Mart has superb processes in store management, inventory, selection, and pricing. GE is world-class in cost management and quality. Toyota is best-in-class in product lifecycle management.

At IBM we know that the product design function—the process by which we decide what products to build, with what attributes and features, at what cost, and at what time to be delivered to the marketplace—is critical in our industry. (This function is also critical, for example, to the automobile industry but not, say, to the petroleum industry.)

Consequently we worked very hard for five years to build a world-class process for product design. It involved millions of dollars of investments, thousands of hours of work, and, eventually, changed the way tens of thousands of IBMers worked. (We have done the same thing with six other processes that we consider crucial to competitive success.)

Great companies cannot be built on processes alone. But believe me, if your company has antiquated, disconnected, slow-moving processes—particularly those that drive success in your industry—you will end up a loser.

Strategic Clarity

Remember the old saying: "If you don't know where you are going, any road will get you there."

No sports team can score if the players don't know what play is called. If everyone has to think about what to do before acting, then confusion and ineptitude are inevitable.

Companies that out-execute their competitors have communicated crystal-clear messages to all their employees: "This is our mission." "This is our strategy." "This is how you carry out your job." But high-caliber execution cannot simply be a matter of exhortation and message. Execution flows naturally and instinctively at great companies, not from procedures and rule books. Manuals may play a role in early training activities, but they have limited value in the heat of battle.

Superb execution is more about values and commitments. At American Express we knew we provided the best customer service in the industry—not because our training manuals said it was important but because our people on the firing line, those who talked to customers all day, believed it. They knew it was a critical component of our success.

The wonderful sales force at The Home Depot who eagerly seek to help you when you visit their stores have a clear understanding of their role in making the company successful. Their behavior emanates from conviction and belief, not from procedures.

On the other hand, too many companies send conflicting signals to their employees. "We want the highest quality in the industry," says the CEO in January. "We need to cut expenses by 15 percent across the board," says the CFO in March. How do the people facing the customer in this enterprise behave the next time a conflict arises over an important customer need?

Mixed signals can be pervasive and difficult. For example, IBM,

I'm sure, always preached the importance of teamwork, yet every-one's pay was based on individual unit performance. We said we value customers above all else, but no one in the field could make a pricing decision without a sign-off from the finance staff.

If you want to out-execute your competitors, you must communicate clear strategies and values, reinforce those values in everything the company does, and allow people the freedom to act, trusting they will execute consistent with the values.

High-Performance Culture

Superb execution is not just about doing the right things. It is about doing the right things faster, better, more often, and more productively than your competitors do. This is hard work. It calls for a commitment from employees that goes way beyond the normal company-employee relationship. It is all about what I call a high-performance culture.

High-performance cultures are harder to define than to recognize. Once you enter a successful culture, you feel it immediately. The company executives are true leaders and self-starters. Employees are committed to the success of the organization. The products are first-rate. Everyone cares about quality. Losing to a competitor—whether it be a big fight or a small one—is a blow that makes people angry. Mediocrity is not tolerated. Excellence is praised, cherished, and re-warded.

In short, businesses with high-performance cultures are winners, and no person of substance would work anywhere else.

[25]

Leadership Is Personal

I deliberately left the subject of personal leadership to last because it is, in my opinion, the most important element of institutional transformation. I mentioned in the chapters on culture that at the end of the day great institutions are the length and shadow of individuals. Great institutions are not managed; they are led. They are not administered; they are driven to ever-increasing levels of accomplishment by individuals who are passionate about winning.

The best leaders create high-performance cultures. They set demanding goals, measure results, and hold people accountable. They are change agents, constantly driving their institutions to adapt and advance faster than their competitors do.

Personal leadership is about visibility—with all members of the institution. Great CEOs roll up their sleeves and tackle problems personally. They don't hide behind staff. They never simply preside over the work of others. They are visible every day with customers, suppliers, and business partners.

Personal leadership is about being both strategic and operational. Show me a business executive who doesn't completely understand the financial underpinnings of his or her business and I'll show you a company whose stock you ought to sell short.

Personal leadership is about communication, openness, and a willingness to speak often and honestly, and with respect for the intelligence of the reader or listener. Leaders don't hide behind corporate double-speak. They don't leave to others the delivery of bad news. They treat every employee as someone who deserves to understand what's going on in the enterprise.

Most of all, personal leadership is about passion. When I think about all the great CEOs I have known—among them Sam Walton of Wal-Mart, Jack Welch of General Electric, Juergen Schrempp of DaimlerChrysler, and Andy Grove of Intel—I know that the common thread among them is that they were or are all passionate about winning. They want to win every day, every hour. They urge their colleagues to win. They loathe losing. And they demand corrections when they don't win. It's not a cold, distant, intellectual exercise. It's *personal.* They care a lot about what they do, what they represent, and how they compete.

Passion. As a student going through Harvard Business School, I would never have guessed that passion would be the single most important element of personal leadership. I don't recall the word ever being spoken during my classroom time at Harvard.

In fact I know I was not sensitive to its role in leadership because of an incident that has stuck in my mind for thirty-seven years. I was interviewing for jobs toward the end of my last year at Harvard. I had narrowed down my search to two companies: McKinsey and Procter & Gamble, the consumer packaged-goods company. At that time, consulting and consumer marketing were considered the two hottest areas in America for MBAs.

The incident took place during my last interview with a very high-level executive at P&G's headquarters in Cincinnati, Ohio. I was an impressionable 23-year-old and had probably never met an executive as senior as this person.

As the interview progressed, I think he sensed my uncertainty (indeed, I was leaning at that time toward consulting). He said something I have never forgotten: "Lou, let's suppose it's Friday night and you are about to leave the office when you get the latest Nielsen report (market-share data for consumer packaged-goods companies). It indicates that you have lost two-tenths of a point of share in the last month in Kentucky. Would you cancel all of your activities for the next day, Saturday, and come to the office to work the problem?"

I remember being startled by the question, and though I didn't give him a definitive answer at the time, the response running through my head was no. I wound up at McKinsey, convincing myself perhaps that I was better off in an environment where the requirements were more "intellectual" and that I would perhaps find it hard to get excited about decimal-point market-share loss of a toothpaste brand.

How wrong I was. As I've stated earlier, a decade later I was frustrated with the detachment and lack of accountability of a consultant. I longed for the opportunity to be responsible for making things happen and *winning, winning, winning*. That senior executive at Procter & Gamble was describing the passion that drives successful executives.

Passion Is for Everyone

All great business executives—CEOs and their subordinates— have passion and show it, live it, and love it. Now, don't get me wrong. I'm not talking about superficial rah-rah optimism or backslapping and glad-handing. Remember my description of personal leadership. It starts with the hard work of strategy, culture, and communications. It includes measurement, accountability, visibility, and active participation in all aspects of the enterprise. Without that, passion is simply a cheerleader doing flips on the sideline while the team gets crushed, 63–0 (maybe 8–0 for those of you who follow soccer).

The passion exhibited by true leaders is not a substitute for good thinking or good people or good execution. Rather, it is the electricity that courses through a well-made machine that makes it run, makes it hum, makes it want to run harder and better.

Exhibiting this kind of passion is a part of every top-notch executive's management style. Who wants to work for a pessimist? Who wants to work for a manager who always sees the glass as half empty? Who wants to work for a manager who is always pointing out the weaknesses in your company or institution? Who wants to work for someone who criticizes and finds fault much quicker than finding excitement or promise? We all love to work for winners and be part of winning. I believe managers at all levels of a company should strive to develop the emotional side of their leadership skills.

I wrote about and listed IBM's Leadership Competencies in the section on culture. One of them was "passion for the business." When IBM's Board of Directors considered who would succeed me, passion was high on their list of necessary attributes. Sam Palmisano, my successor, is an extraordinary executive—a man of many talents. However, he would never have had my recommendation, despite these many talents, if he didn't have a deep passion for IBM, for what it stands for, for what it can be, for what it can do. He has an emotional, 24-hour-a-day attachment to winning and to achieving ever-increasing levels of success.

WHAT IT TAKES TO RUN IBM

Energy

- Enormous personal energy
- Stamina
- Strong bias for action

Organizational Leadership

- Strategic sense
- Ability to motivate and energize others

- Infectious enthusiasm to maximize the organization's potential
- Builds strong team
- Gets the best from others

Marketplace Leadership

- Outstanding oral communications
- CEO-level presence and participation in the industry and with customers

Personal Qualities

- Smart
- Self-confident, but knows what he/she doesn't know
- Listens
- Makes hard decisions—in business and with people
- Passion that is visible
- Maniacal customer focus
- Instinctive drive for speed/impact

Integrity

I want to close this chapter on personal leadership with a few comments about integrity. All of the great leaders I have known may be tough (in fact, all of them were tough-minded, which is very different from some people's description of "tough"). However, all of them were, at the same time, fair. Fairness or even-handedness is critical for successful leadership. Playing favorites, excusing some while others hang for the same offense, destroys the morale and respect of colleagues.

This concept sounds simple, but is very hard to carry out every day. I could not begin to count the number of times during my decade at IBM when an executive would appeal to me for an exception to our principles or policies. "John didn't make his numbers this year, but he tried very hard. I think we should still pay him a good bonus so that he stays motivated and doesn't leave." "Susan got an offer from a competitor and I know that if we match it we will upset the compensation scheme in the finance function, but we have to make an exception to keep her." "I know it looks like Carl was involved in a sexual-harassment incident and we have fired others in similar circumstances in the past, but Carl is too critical to the success of Project X. He's very apologetic and will never do it again. So let's just slap him hard but not fire him."

In hundreds of such conversations, there were always two sides to the story; there was always a seemingly good reason to bend the rules and make exceptions. And, examined one by one, in every case the executive can talk himself or herself into making an exception.

Cumulatively, however, if an executive demonstrates that exceptions are part of the game, then his or her leadership will erode as the trust of colleagues evaporates. Cultures in which it is easier to ask forgiveness than permission disintegrate over time. Leaders who don't demand uniform and fair adherence to good principles and policies lose their effectiveness.

Postscript

This chapter originally ended here. However, with all the news of corporate malfeasance that has emerged in mid-2002, I need to add a postscript. My preceding comments deal with the inevitable challenges that all leaders face to maintain an environment of fairness and principled judgment. I did not think it was necessary here to deal with dishonesty and law-breaking, or with lying and stealing.

No one should be entrusted to lead any business or institution unless he or she has impeccable personal integrity. What's more, top-rung executives have to ensure that the organizations they lead are committed to a strict code of conduct. This is not merely good corporate hygiene. It requires management discipline and putting in place checks and balances to ensure compliance.

If any of these allegations about certain executives turns out to be true, this is simply unacceptable behavior by bad people. I'm ashamed of them and embarrassed by them. They are, however, a very small subset of the corporate world. I believe the vast majority of our business leaders are good, hard-working people who live up to the standards of integrity that we expect of all those whom we entrust with power and authority.

26

Elephants *Can* Dance

For much of my business career, it has been dogma that small is beautiful and big is bad. The prevailing wisdom has been that small companies are fast, entrepreneurial, responsive, and effective. Large companies are slow, bureaucratic, unresponsive, and ineffective.

This is pure nonsense. I have never seen a small company that did not want to become a big company. I have never seen a small company that didn't look with envy on the research and marketing budgets of larger competitors or on the size and reach of their sales forces. Of course, in public, small companies put forth David v. Goliath bravado, but in private they say, "I wish I could work with the resources those big *!#@* have!"

Big matters. Size can be leveraged. Breadth and depth allow for greater investment, greater risk taking, and longer patience for future payoff.

It isn't a question of whether elephants can prevail over ants. It's a question of whether a particular elephant can dance. If it can, the ants must leave the dance floor.

I don't intend to describe here all the elements of creating a nimble, responsive, large enterprise.[1] Certainly the matters just discussed—focus, execution, and leadership—apply to enterprises of all sizes.

There is one item, however, that I want to comment on, because it was essential in getting IBM dancing again. This is the issue of centralization v. decentralization in large enterprises.

A corollary of the "small is good, big is bad" mantra is the popular notion that, in large enterprises, decentralization is good and centralization is bad.

In the 1960s and 1970s McKinsey built a powerful reputation promulgating decentralization to corporations all over the world. It first pushed the idea in the United States, moved into Europe in the 1970s, and eventually went to Japan (where the idea was rejected emphatically by most Japanese companies).

Decentralization had a powerful intellectual underpinning, and over the course of a few decades it became the "theory of the case" in almost every industrial and financial enterprise. The theory was very simple: "Move decision making closer to the customer to serve that customer better. Give decentralized managers control over everything they do so they can make decisions more quickly. Centralization is bad because it inevitably leads to slower decision making and second-guessing of the people on the firing line, closest to the customer. Big companies are inevitably slow and cumbersome; small companies are quick and responsive. Therefore, break big companies into the smallest pieces possible."

There's a lot to be said about the power of this construct, and it should, in my opinion, continue to play an important role in organizational behavior in large enterprises. However, I believe that in the 1980s and 1990s it was carried to an extreme in many companies, with

---

1   This issue of entrepreneurial behavior in large corporations had been a passion of mine for decades. See Harvard Business School cases on corporate entrepreneurship, based on my activities at American Express. (Harvard Business School Cases 9-485-174 and 9-485-176, copyright © 1985).

unproductive and, in many cases, highly disruptive results. Too often managers began to express the view that they lost their manhood or womanhood if they didn't control *everything* that touched on their business. Consequently, every decentralized business had its own data processing center, human resources group, financial analysis team, planning organization, and so on. Decision making was, in fact, fast if the decision touched only on a single decentralized unit. However, when multiple segments of the enterprise had to be involved, the highly decentralized model led to turf battles and inadequate customer responses because of incompatible systems in the bits and pieces of the enterprise.

Moreover, as long as profit margins were fat, the extra staffing may have been tolerable, but as we approached the raw-knuckled competition of the 1990s with capacity excesses in almost every industry, companies could ill afford duplicating staffs and process development at every level of the company.

Yet cost and speed are not the only issues. In many large institutions, the decentralized units were created for a different world or acquired as pieces of a larger mosaic. Now these companies are trying to create new value through the combination of historically separate entities. Examples abound all over the corporate world: financial services companies striving to create integrated offerings for customers from historically disparate product units; industrial companies trying to redefine their value to their customers as something more than a traditional product—usually a service wraparound; media companies trying to package advertising opportunities that combine various pieces of their enterprises; telecommunications companies trying to attract and hold customers through integrated offerings of voice, data, and entertainment.

This is not a challenge limited to the corporate world. University presidents have been struggling for decades to create interdepartmental programs that integrate various fiefdoms of the academy. Memorial Sloan-Kettering Cancer Center has been working for years to create cross-departmental treatment protocols, i.e., an integrated approach to a particular type of cancer that combines surgery, chemotherapy, and radiology. Both in universities and medical centers this is hard work, because the department chairs who run the traditional decentralized units have enjoyed years of carefully guarded autonomy.

The problem of decentralization exists in government, too. The United States intelligence community is a hopeless hodgepodge of overlapping yet ferociously independent organizations. When a new threat arises (such as domestic terrorism), the task of redirecting the intelligence assets of the country away from the missions they were originally designed to carry out to meet the new challenge becomes an integration task of gigantic proportions.

Too Expensive, Too Slow

I believe that in today's highly competitive, rapidly changing world, few if any large enterprises can pursue a strategy of total decentralization. It is simply too expensive and too slow when significant changes have to be made in the enterprise. Thus, what every CEO has to do is decide what is going to be uniquely local (decentralized) and what is going to be common in his or her enterprise. Note the absence of the word "centralized." It is not a question of centralization v. decentralization. Great institutions balance common shared activities with highly localized, unique activities.

Shared activities usually fall into three categories. The first and easiest category involves leveraging the size of the enterprise. Included here would be unifying functions like data processing, data and voice networks, purchasing and basic HR systems, and real estate management. For the most part these are back-office functions that yield to economies of scale. It is absolutely foolish for a CEO to accept the whining of a division president who says, "I can't run my business successfully without running my own data center, managing my own real estate, or purchasing my own supplies." Even a company as diverse as General Electric effectively exploits its scale economics in back-office processes.

The second category involves business processes that are more closely linked to the marketplace and the customer. Here the drive to common systems can offer powerful benefits but most often involves linkages among the parts of a business that may or may not make sense.

I'm thinking here of common customer databases, common fulfillment systems, common parts numbering systems, and common customer relationship management systems that permit your customer-service people to provide integrated information about everything a customer does with your company.

On the surface it would seem that these are logical and powerful things to do in an enterprise. Nevertheless, they usually require profit-center managers to do something very hard—relinquish some of the control they have over how they run their business. Staff executives, consultants, or reengineering teams cannot do this without active line management involvement. The CEO and top management have got to be deeply involved, reach tough-minded conclusions, then ensure that those decisions are enforced and executed across the enterprise. It takes guts, it takes time, and it takes superb execution.

A Step Too Far

Having made the point that decentralization has gone too far in many institutions, I quickly add that there is a ditch on both sides of the road. My concern is that today many CEOs are seeking utopian levels of integration. This is the third—and most difficult—area of common activities, involving a shared approach to winning a marketplace, usually a new or redefined marketplace. These activities are

difficult because they almost always demand that profit-center managers subjugate their own objectives for the greater good of the enterprise. As such, they can be enormously controversial inside a company
and lead to bitter and protracted struggles.

Here's an example: During my time at American Express I was
running the so-called Travel Related Services business, which included the American Express Card division. It was the largest and
most profitable segment of American Express. American Express
bought a brokerage company as a step to create a one-stop financial
supermarket. In the course of enticing the brokerage company to join
American Express, the deal makers promised the brokerage that they
would have access to the American Express cardmember list. In other
words, they would be allowed to make cold calls to cardmembers to
try to sign them up for brokerage accounts. When this became known
to the card division, there was an open revolt. Those of us who had
built the card division believed it was assembled on a basis of trust,
privacy, and personalized service. Cold calls from securities brokers
did not fit into our definition of customer service.

The war went on for years, and the integration or synergy that
the CEO had hoped to achieve not only never happened, but it led to
the departure of many senior executives and ill will that contributed
to the eventual disposition of the brokerage business.

It is very easy to conceive of how various units in a company can
work together against a common enemy or seize new ground in a
competitive industry. Think about all the financial supermarkets that
have been constructed (and almost as many deconstructed). Think
about all the mergers and acquisitions that have taken place in the entertainment and media industries; The New York Times Company
buying cable companies and sports magazines; Disney buying a television network; the behemoth known as AOL Time Warner.

How many times have we watched two CEOs stand up at a press
conference and make claims about the extraordinary benefits that
would be achieved once they merged their companies to create a

unique combination that would bring new services and new benefits to the marketplace?

Well, we've all seen what happened in almost every one of these instances. They fail. Why? Because in most cases the CEO must ask people to do things that are inextricably and inexorably in conflict. Divisions are asked to compete against their traditional competitors, focusing on maintaining a leadership position in their individual markets. At the same time, they're asked to join with other divisions in their company in a much broader fight that inevitably involves giving up some resources or assets that are needed to win in their traditional market.

There is great risk in asking a decentralized unit of an enterprise to be good at its traditional mission and, at the same time, fulfill a shared role in creating value in a new mission. The conflicts—most often having to do with resource allocations, but also with pricing, branding, and distribution—will be overwhelming.

I am about to suggest something that will annoy almost all the world's management consultants (they make a lot of money defining "new industry models" and describing "synergy opportunities"): CEOs should not go to this third level of integration *unless it is absolutely necessary.*

For most enterprises the case for integration ends with category two. Category one is a no-brainer; most back-office functions can be combined with significant economics of scale. Category two (integration of "front office" functions that touch the marketplace) can produce significant benefits, but the integration must be executed superbly or the benefits will be decimated by the parochial interests of individual units. Category three is very much a bet-the-company proposition.

However, there will be times when a CEO feels it is absolutely necessary to bet the company on a new model—a truly integrated model (I recently met on two occasions with CEOs who run media and entertainment companies, and they said they are agonizing over this decision). If you decide to go down this path (as I did at IBM), let me outline a few of the steps that I think are critical to making a successful conversion. I can't and won't belabor this issue here; it would fill another volume. What follows are only very introductory comments.

Shift the Power

One of the most surprising (and depressing) things I have learned about large organizations is the extent to which individual parts of an enterprise behave in an unsupportive and competitive way toward other parts of the organization. It is not isolated or aberrant behavior. It exists everywhere—in companies, universities, and certainly in governments. Individuals and departments (agencies, faculties, whatever they are called) jealously protect their prerogatives, their autonomy, and their turf.

Consequently, if a leader wants fundamentally to shift the focus of an institution, he or she must take power away from the existing "barons" and bestow it publicly on the new barons. Admonishments of "play together, children" sometimes work on the playground; they never work in a large enterprise.

At IBM, to be a truly integrated company, we needed to organize our resources around customers, not products or geographies. However, the geographic and product chieftains "owned" all of the resources. Nothing would have changed (except polite platitudes and timely head nodding) if we didn't redirect the levers of power. This meant making changes in who controlled the budgets, who signed off on employees' salary increases and bonuses, and who made the final decisions on pricing and investments. We virtually ripped this power from the hands of some and gave it to others.

If a CEO thinks he or she is redirecting or reintegrating an enterprise but doesn't distribute the basic levels of power (in effect, redefining who "calls the shots"), the CEO is trying to push string up a hill.

The media companies are a good example. If a CEO wants to build a truly integrated platform for digital services in the home, he or she cannot let the music division or movie division cling to its existing technology or industry structure—despite the fact that these traditional approaches maximize short-term profits.

Measure (and Reward) the Future— Not the Past

I have already pointed out that people do what you *inspect,* not what you *expect.* Leaders who are thinking about creating true integration in their institutions must change the measurement and reward systems to reinforce this new direction.

I recall one of the senior executives at American Express who was big on synergy. He spoke about it constantly. Yet, in all the financial reports the total focus was on the traditional, independent profit centers. I remember the poor souls who were assigned the task of creating synergies among the card, Traveler's Cheque, and travel divisions. At best they were tolerated; more often they were simply ignored. The compensation system at American Express did not help—98-plus percent of a manager's annual compensation came from the results of his or her specific unit. "Synergy points" (which became an exercise in creative writing around bonus time) might add a minuscule amount to your pay.

I knew we could not get the integration we needed at IBM without introducing massive changes to the measurement and compensation system. I've already explained that the group executives who ran IBM's operating businesses were not paid bonuses based on their unit's performance. All their pay was derived from IBM's total results.

When a CEO tells me that he or she is considering a major reintegration of his or her company, I try to say, politely, "If you are not prepared to manage your compensation this way, you probably should not proceed."

Measuring financial results is the same issue. We were never able to move to an integrated customer view, rather than a geographic view, until we stopped creating P&L statements for the geographic units. Of course, many of the geographic leaders went ballistic! "We can't manage our business without P&L oversight." "Sorry," I said, "you no longer manage a business. You now serve as a critical support function to our integrated worldwide customer organization."

Walk the Talk

As with much that I have discussed in this book, CEO leadership is mandatory before substantial changes become systemic and sustainable: They require real involvement and not exhortation, delegation, and then surprise when change doesn't happen.

It took me more than five years of daily attention to get IBM to accept a new go-to-market model. It was a tremendous battle. If you choose to follow a similar path, you must be prepared to make it happen personally! The assignment cannot be delegated. Who would you delegate it to? The operating team that hates the loss of autonomy? Staff executives who will be ambushed and disemboweled by those fated to lose power? No. It's a lonely battle, and it explains why, after twenty years of talking:

- There are *no* true financial supermarkets.
- There are *no* integrated, multi-service telecommunications companies.
- There are *no* fully integrated entertainment companies.
- There are *plenty* of financial services companies spinning off their insurance and/or money management businesses.

- There are *plenty* of divestitures of cable systems and wireless assets by telephone companies.
- There is *plenty* of skepticism about "convergence" in the entertainment industry.

Much of the press coverage of IBM over the past decades was focused on our strategic restructuring—as well it should have been, since without that restructuring there would be no IBM today. However, as I pointed out earlier, our current strategies will—and should—change as the industry continues to evolve very rapidly. History determines legacies, but if I had a vote, the most significant legacy of my tenure at IBM would be the truly integrated entity that has been created. It certainly was the most difficult and risky change I made.

PART V

Observations

In my thirty-five-year business career, I have observed the world from many viewpoints. I have ridden the ups and downs of economic, industry, and product cycles. I have introduced new products, revived ailing old ones, and shut down longtime flops and launch-pad failures. I have walked on all but one continent. (IBM does not sell a lot of computers in Antarctica!) I have been the guest of honor at business dinners at which the main course was scaly anteater or bear paw or camel hump, and I ate everything set before me without wincing or gulping because it wasn't proper to offend my hosts.

And, as I write this, I am 60 years old. That's not as old as it used to be, but it's well past halftime.

A long time ago I promised myself that I would never become a cranky old man complaining to waiters and clerks, yelling at the neighbors, and honking the car horn at everyone who drives a mile over the speed limit. Yet, as I think about this final section of the book—observations on a variety of things after ten years at IBM—I am staring right into the eye of temptation to air a long list of annoyances.

I'm going to resist that temptation as best I can—but not completely. I want to make it crystal clear, however, that what follows are only personal observations. They are not necessarily shared by my colleagues at IBM.

27

The Industry

Having worked in more than half a dozen industries in the course of my career as a consultant and an operating executive, I think I'm qualified to say that the information technology industry is truly unique.

When I got to IBM, I was prepared to be challenged by the considerable technology that drives the industry. What I was totally unprepared for, however, were the characters and bizarre practices of the computer industry.

Let's begin with some of the industry leaders. It is a truly extraordinary cast: Larry Ellison of Oracle, Scott McNealy of Sun Microsystems, Bill Gates and Steve Ballmer of Microsoft, and Steve Jobs of Apple. These men obviously possess enormous talent. They have built important, ongoing enterprises. Each one of them qualifies many times over for my important criterion of having passion for the business. However, they are also among the most outspoken leaders I have ever met. They make outrageous remarks, they attack one another publicly with great relish and they seem to have no qualms about denigrating the other guys' products, promises, and pronouncements.

It's a 24/7/365 three-ring circus. I'm not pointing fingers at any individual, but I have to say, I've never witnessed any such behavior in all the other industries where I've worked. If I'd planned to say even a tenth of what passes for normal business discourse in this industry, my lawyers would have bound and gagged me and bolted the door.

Of course, there are more traditional, successful CEOs like Michael Dell and Andy Grove. And, of course, the industry also contains thousands of CEOs of very small companies who, for the most part, are dedicated technical people trying to create a successful niche and at the same time not get trampled by the giants or eaten by the piranhas.

Interestingly, there is no trade association for the IT industry. Other industries have such associations—like the American Bankers Association, the Grocery Manufacturers Association, the Pharmaceutical Industry Manufacturers Association. There are some groups that represent segments of the IT industry and other groups that come together periodically to address important issues.

A cynical person might conclude that values like teamwork, common causes, and mutual respect are not in the industry's DNA. Having attended banking and grocery industry meetings and having seen the conviviality and personal relationships expressed among these CEOs, I would love to be a fly on the wall if we were ever able to bring all the IT masters of the universe together in one room!

The second thing I find remarkable about this industry is the nature of the competitive battles. They are truly ferocious struggles where the objective is not simply to increase your market share by a few points, but also to annihilate your competitors. This is not just bravado. It's a strategic imperative that is unique, in my opinion, to the computer industry. Most industries follow the law of diminishing returns—i.e., after a certain point it costs more to increase your market share against entrenched competitors than you get in return. As a result, most industries arrive at an equilibrium among three to five competitors and share movement becomes very difficult to achieve (unless one company really falls apart).

The computer industry, however, dances to a different drummer. It follows a rule of "increasing returns," in which it is possible to become *the* standard. And when you become the standard, you effectively "own" a market. This is related to a "networking effect," which argues that the more people there are on your network, the more valuable and profitable your network becomes and the less viable is the competitors' offering. As a result, competitive battles are truly vicious in this industry. The objective is always to "sweep the floor" in a customer account and drive the incumbent or shared competitors off the premises.

The third aspect of the industry that I find amazing is its absurd preoccupation with underlying technology. IT companies really believe the wild pronouncements they make. Everyone is searching, all the time, for the next big wave.

Think about all the guaranteed, fail-safe, revolutionary predictions in our lifetime—the cashless society, the paperless office, the PC revolution, and the dot-com universe (you'll never need to leave your home to shop, go to school, do your banking, see the world). We have been told that computers will respond easily and automatically to speech commands. It's all breathtakingly appealing at times.

But is it true? Does it really happen? We all still have cash in our pockets. We're still writing more checks than when the checkless society was proclaimed. You can't go into an office anywhere in the world and not see piles of paper. The PC remains a difficult, clunky, and inefficient device. The dot-coms are largely gone. And the sun still comes up every morning and life goes on.

There are many reasons why these grand pronouncements never work out, but far and away the most important one is the remarkable detachment that the industry maintains from its customers. Of course it makes sense to carry out all financial transactions online, without checks. Of course it would be great if we had no paper in our offices. However, these concepts ignore human behavior, human preferences, human biases, and institutional and personal demands that emanate from the nontechnical parts of people's and companies' lives. Somehow people in the technology industry think that every one of their customers wakes up every morning saying, "I wish I had more technology. I can't wait to learn more about what computers can do."

I just wish every one of these incredibly bright technologists could spend a year as a customer and see the different viewpoint customers have about computing technology. They would see that customers find technology very difficult to integrate into everyday lives and enterprises. They would find the promises overblown and the returns more difficult than promised. They would find that, at the end of the day, many of the critical decisions that managers, employees, and consumers have to make either have no relationship to technology, or they just may find that technology can actually be an impediment.

Having read all this, you may be surprised when I say that I loved my decade in the IT industry. While the aberrations I've described from normal business practice are huge and irrational, you cannot find a more exciting place. If you love competition, if you love winning, if you love change, if you love building exciting new things, if you love the intellectual stimulation as well as the emotionally draining, seven-days-a-week commitments, there's no better place. I may find Scott McNealy outrageous, but I also find him fun to be with and always provocative and a worthy competitor.

28

The System

The American economy is currently nursing a bad hangover. Investors have lost trillions of dollars in the last several years. MBAs are dazed as they plummet from multimillionaire to unemployed.

Most disturbing of all, a handful of CEOs and CFOs have betrayed the trust and loyalty of their shareholders and employees. They should be held accountable to the full extent. Unfortunately, while their punishment will alleviate some of our shock and anger, it will not replace the jobs and personal savings of thousands of Americans. The financial scandals of 2002 have caused considerable damage to the reputation of the entire business community.

Nevertheless, we must be careful in our response. Every institution—corporations, governments, the federal bench—has people who cheat, take advantage, lie, and play loose with the rules. However, nowhere are they in the majority. We need to punish the criminals without weakening the institutions that are critical to our economic and social well-being.

Regulators and politicians are conducting a nationwide search for anyone even remotely involved in the crashes of onetime high-flying businesses. Media companies that had never—or rarely—

reported stories about corporate accounting are now whipping a horse that was proudly displayed by these same media companies only a few years ago. CEOs and CFOs of dot-coms and, of course, telecom, media, and energy companies had appeared on the financial news stations to boast about their pro forma earnings. The commentators reveled in the coverage because everyone wanted more news of the "new economy." The print and broadcast media lionized the new economy and rarely cast even a bit of doubt on companies whose stocks were selling at 300 or 400 times their revenue!

What I find amazing is not that all of this happened but that few said we had seen this movie before and knew the climax was going to be a big train wreck. Whether it was the California gold rush, or the wild industrial expansion of the 1920s, or the conglomeration of the 1960s, or the LBO phase of the 1980s, our wonderful and free economic system breeds periods of wild enthusiasm and speculation. There is no central bad guy in these movies; everybody's a bad guy. Entrepreneurs overpromise, corporate executives get greedy, commercial bankers forget the basic rules of credit policy, politicians kowtow to the new titans of industry (remember all those treks to Silicon Valley during the last two presidential elections?), and the financial media race to the bank with burgeoning advertising revenue and a heady mix of boosterism to keep the drumbeats going.[1]

While most everyone imbibes from the punch bowl, there still had to be somebody who arranges the party and supplies the booze. The suspects being hauled into court right now—accountants, regulators, and corporate executives—are the last people in the world who could pull off a blockbuster binge like the one we've just completed.

Who could pull it off? Look no further than investment bankers. They supplied the hooch for all the wild speculative periods in our economic history. They created the great oil cartels. They sold

---

1   For an interesting study of the events of this era, read *Dot-con,* by John Cassidy, HarperCollins, February 2002.

the railroad stocks. They worked beside the great conglomerateurs of the 1960s. They then turned around and, for another set of fees, sold off all the businesses that made no sense to conglomerate in the first place. They have an infinite variety of matches to start economic bonfires: EBITDA, IPOs, LBOs, MBOs, PERCs, and PICKs.

They make money coming and going. They make huge fees telling AT&T to buy up everything in sight, then make more huge fees helping AT&T to sell off everything. All the deal makers are rich at the end of a speculative binge. The losers are the investors, who put up their hard-earned money; and the entrepreneurs, who put up their talent and reputation.

So what do we do to fix the system? Some of you are going to hate this answer: not much.

Closing down the investment bankers is a nonstarter. Remember, they exist (and will always exist) because people are always seeking a magic potion to fulfill their dreams. We need little new regulation. Regulation usually strangles economies; it rarely supports them.

Some of the ideas now being discussed make sense—for example, the Securities & Exchange Commission ought to ban the use of pro forma earnings in corporate press releases. Boards of Directors should be composed almost entirely of outside directors. Employees should not be required to keep the major portion of their employee retirement and savings assets in their company's stock.

I do think there's one significant change that would be helpful in dampening the next wave (and, believe me, we will have another wave in the next ten or fifteen years). My suggestion—it is not original with me; it has been made by many others—is to change our tax policy to discourage short-term speculation and encourage long-term investment. I would like to see profits made from the buying and selling of securities taxed at, say, 70 percent if the transaction takes place in less than a year; 40 percent for one to three years; 20 percent for three to five years; and zero for anything over five years. This would help the small-business people, as well as the family farmer, and encourage corporate executives to think beyond quarter-to-quarter performance.

I would apply the same approach to stock-option gains. Executives who exercise options and sell the stock in less than five years would pay substantially higher taxes than those who exercise options and hold the stock, which is the only reason to give stock options to executives—i.e., to align their interests with long-term shareholders.

One of the major benefits of this change in tax treatment would be, I hope, a return to "owner capitalism." If large-equity stakeholders (pension funds, mutual funds) are severely penalized for short-term trading, and concurrently rewarded for long-term investing, they might become constructive overseers of corporate performance.

Unfortunately, we have been going in the opposite direction in the United States over the past decade or so. United States regulators have become so preoccupied with helping the small investor that, I'm sure unwittingly, they have disconnected the large investors—the owners most able to exert pressure on the board and management.

For example, shareholders' meetings—the once-a-year opportunity for investors to interact publicly with managers—have been turned into a circus by the regulators (this is not only true in the United States but in Europe and Japan, as well). I remember, twenty years ago, being part of a management team that prepared very diligently to address important issues about our business at the annual shareholders' meeting. Today no serious institutional shareholders go to annual meetings. These meetings have become, because the regulators have permitted it, showboating opportunities for gadflies, social engineers, and disgruntled employees. I'll never forget an annual meeting of AT&T when shareholders were forced to sit through long harangues—twice—for and against abortion, because regulators had permitted two abortion-related proposals. All this crowded out any discussion of critical issues that should have been addressed (for example, what was AT&T doing to improve margins in the long-distance business?).

Another example of good intentions gone awry is the new requirement in the United States known as regulation for fair disclosure (FD). It purports to ensure that all investors—especially small ones—get access to corporate information at the same time. Nobody can argue with that objective.

However, it appears that it can also induce management to shun the deep reviews with large institutional investors that, in the past, were the sternest tests for corporate leaders. Believe me, if you were a CEO or CFO preparing for a two- to three-hour meeting with Fidelity or American Express Financial Advisors, you would focus on explaining and defending your stewardship of the company in a tough, give-and-take environment. I'm afraid much of this may disappear under FD, if it hasn't already.

My point here is simply that it is the large, sophisticated owners who can and should be the principal force for holding management accountable. Yes, regulators, directors, and accountants play a role. But we should take a page from the venture capitalists' book: Owners who are close to the managers, owners who cannot sell their stock on a whim, owners who must see a company through a complete investment cycle—they are the most powerful force for ensuring effective corporate leadership.

29

The Watchers

I've had a sign in my office for many years. It says:

THERE ARE FOUR KINDS OF PEOPLE:

THOSE WHO MAKE THINGS HAPPEN.

THOSE TO WHOM THINGS HAPPEN.

THOSE WHO WATCH THINGS HAPPEN.

THOSE WHO DON'T EVEN KNOW THINGS ARE HAPPENING.

This book is all about IBMers who made things happen.

The group to whom things happen is a complex lot. Some are happy followers—go-with-the-flow people. Others are resisters to change and they, too, have been discussed at length.

I envy the fourth group—the people I have met from all walks of life who go through their days absorbed in their own world and are spared the agony, fear and frustration of current events, social change, and the great issues of the moment.

It is the third group I want to discuss with you now—those who watch things happen.

For the most part watchers don't create. They observe and comment on the work of others. (You might call them professional opinion-havers.) Consequently, their value in society is a function of the insightfulness of their observations and the extent to which their comments add value to a process in which they are fundamentally outsiders.

I wish to emphasize that "value in society" is not the same as "success." I believe there are many watchers who add little value but who are successful because they create controversy, stir emotions, entertain, or appear on TV a lot. On the other hand, few things are as rewarding as reading a truly insightful magazine article about, say, the genome project or the antecedents of the Middle East crisis.

As CEO of one of the world's largest companies, I wanted to spend every moment possible worrying about customers, employees, and shareholders. However, I knew that I couldn't escape dealing with the IBM watchers, because if they decided for whatever reason that they didn't like me or what I was doing, my job was going to be even more difficult.

There are many categories of watchers. Talking about them—watching the watchers—can be risky because they usually have the last word. That's why I'm going to limit my discussion to the most prominent watchers—the business media and analysts—and I will muster as much objectivity as I can.

The Business Media

As in every other profession, there are good reporters who cover business for newspapers, magazines, and TV, and there are not-so-good ones. I have been fortunate to work with some of the best—reporters who took the time to get to know the companies and industries they covered and who asked insightful questions and wrote stories with impact.

Outside the United States, I was most impressed by reporters from Japan. They were very thorough—at one press conference I held in Tokyo, a large newspaper sent sixteen reporters. They worked hard on their stories, and I found the reporting to be almost always accurate. Japanese reporters may well be the most aggressive in the world. They think nothing of tracking you down at your hotel room and knocking on the door late at night. Our head of IBM Japan told me that reporters frequently knock on his door at home at midnight—and he invites them in for a beer!

I have also had to deal with—or live with the consequences of—some not-so-good reporters. They have included these types:

- Those who believe that only bad news is news. One such IBM beat reporter trolled Internet chat rooms every day, searching for disgruntled former employees and calling short-sellers, who like nothing better than to have negative rumors spreading through the markets. If a story this reporter wrote drove down our stock, the next day the publication would report accurately that the stock went down because of a story in that publication. Most people would call that a twofer. Another beat reporter at another publication listed his name and phone number on Internet gripe sites for people who had negative observations and experiences concerning IBM.

- Those who chase headlines at the expense of content. You could always tell during an interview. The reporter keeps prodding, poking, trying to get you to say something controversial, sometimes asking the same question three or four times. For the most part, these reporters didn't care about or understand what was going on at IBM.

- Those who let you believe they are writing a story with one kind of slant, when in fact they plan a totally different angle. I can't recall that any of these reporters ever told a bald-faced lie. But trickery and illusion certainly are not synonymous with honesty.

- Those who simply dislike you. One of our beat reporters went on national TV and said that the people at the top of IBM were bad guys.

Later when he called to say he planned to write a story about us and wanted to assure us that he had an open mind, what were we to think?

Whenever it became apparent that a reporter was one of these not-so-good types, I had a simple policy: I refused to deal with him or her. Life is too short. And I had a lot of other, more productive things I could do with my time.

Underpromise and Outperform

I took a lot of heat from the media on one issue throughout my nine years at IBM: access to the media. Some of my competitors seemed to be unable to walk past a TV camera or a reporter. I had no such problem. The complaints were loudest from the technology trade papers. "Bill Gates personally answers his phone when I call. Why can't Lou Gerstner?"

Well, I had a company to run. More important, I have always believed that it is better to underpromise and outperform than to overpromise and underperform. The media like forecasts, predictions, and promises. If you spend a lot of time on camera, you are likely to succumb to the temptation to overpromise.

So throughout my business career it was my practice to do only two interviews—maybe three—a year. Interviews were almost always part of a major story by a major publication, and usually with a reporter who had a story line that demonstrated an insightful view of industry trends and a deep understanding of the company's strategic directions. When I was given the choice of seeing a reporter who had such a specific story line or one who simply wanted to ask a lot of questions about a plethora of different subjects, I almost always picked the reporter who had a specific story line.

I've always believed that less is more when it comes to the press.

I may have limited my access to the media, but I'd argue that the few interviews I did do had greater impact than if I had worried about a daily sound bite.

Analysts

Analysts come in lots of flavors: observers who specialize in opinions on stock prices, on industries, on products. And, as is the case with all watchers, some analysts are very good, some not so good. I have already commented on the Internet analysts, many of whom sold their clients down the river.

We have some good securities analysts who follow IBM. One in particular impressed me recently when he submitted a list of questions he wanted answered at an upcoming IBM analysts' meeting. They were good strategic observations.

Most of the security analysts I've known are hard-working, bright people. They suffer, however, from a very debilitating virus that has infected our public capital markets in America: quarterly myopia. It is amazing to me how many analysts have lost sight of the forest for the trees. Or, perhaps stated differently, they have divided the forest into a complex maze—like what you see on old English estates. All that matters to these analysts is that you make the next short-term passage without a falter. It really doesn't matter that you have the best map to get through the total passage (or that you have no map!). If they predict 50 cents a share in the next 90 days and you make it—you're a hero. If you make 48 cents, you're a goat or a villain.

Never mind that 90 days has absolutely no bearing on how well a company is doing. Never mind that in most businesses there is little a manager can do to effect real results in 90 days. Never mind that what a company *can* do in 90 days is manage earnings. In defense of these security analysts, I concede that they did not create this myopia; they are part of a much wider epidemic. Yet the system turns them into spreadsheet junkies, hoping every quarter you will fill their bingo card with the right numbers.

There is one misguided notion, however, that I do lay at the feet of analysts: a sophomoric preoccupation with revenue growth as a measure of a company's strength. Of course increasing revenue is an element in creating value in a corporation, but it is far from the most important factor. Chasing revenue at the expense of real earnings is one of the most telling signs of a weak management team. Of all the measures on a profit-and-loss statement, revenue is one of the easiest to manage through creative bookkeeping. We need look no further than the myriad of telecom, dot-com, and software companies that are being investigated and/or are confessing to such manipulations as this book goes to press.

A preoccupation with revenue can also lead to maximizing short-term results at the expense of long-term competitive position, e.g., raising or holding prices at the expense of building market share.

There are many things we did—or avoided—at IBM that depressed revenue growth but increased fundamental 'shareholder value:

- We did not make acquisitions that added size but little profit.
- We reduced mainframe prices and, therefore, revenue to ensure a powerful ongoing stream of cash flow.
- We undercut the prices of some of our software and storage competitors that were maintaining outrageous price umbrellas over their products. As a result, we added huge market-share gains in high-growth areas.
- We sold businesses that generated billions of dollars of revenue but were not strategic for IBM in the long term.

I presume revenue is something analysts find easy to track and understand. However, the best companies grow profits faster than revenue. They manage margins and expenses brilliantly. They know that the best competitive advantage is a cost structure and a go-to-market model that allows them to use revenue as a weapon against their competitors. Most important, they understand that cash flow, not revenue, sustains corporate success.

It should not surprise anyone that three of the IT companies that have grown the strongest over the last decade—Dell, Intel, and IBM—share two common traits (traits they also share with GE, Wal-Mart, and other leaders). First, they manage pricing (revenue) as a strategic competitive weapon; second, they manage costs superbly well—all costs, including taxes, manufacturing, and distribution.

As we proceed through this new decade, the leaders will use e-business capabilities to create new innovations in manufacturing, engineering, and supply chain, and in human resource productivity and organization speed.

In almost every industry, globalization is leading to overcapacity, which is leading to commoditization and/or price deflation. Success, therefore, will go to the fittest—not necessarily the biggest. Innovation in process—how things get done in an enterprise—will be as important as innovation in the products a company sells.

If I were the head of research at a securities firm, I would urge my analysts to focus on the following five points in determining shareholder value:

1. Is the company a major force in a growing market or market segments? (Remember Warren Buffett's wonderful observation: "When a manager with a great reputation meets a company with a bad reputation, it is the company whose reputation stays intact.")

2. Is the company holding or increasing its share in those segments, and is that share gain the result of sustainable advantages (cost, technology, quality)?

3. Is the increased share resulting in growing cash flow—cash flow after *all* expenses, not the notorious EBITDA (earnings before interest, taxes, depreciation and amortization), and not pro forma nonsense?

4. Is the company using that cash flow in a wise manner?
   - avoiding macho or bleary-eyed acquisitions
   - reinvesting in research and development, marketing, and other critical areas in the company

5. Does the management team walk the talk of aligning with the shareholders? Do executives own significant amounts of stock (as opposed to just holding options)? Do they return cash to their shareholders in the form of dividends or share repurchases?

I love competitors who get hooked on revenue as a key target of their performance. I was gleeful in 1993 when I heard the head of Compaq announce that his goal was to surpass IBM in revenue by 1996. The analysts cheered! He went off and bought Digital Equipment, and I cheered! He was eventually fired, and Compaq has since disappeared as a corporation.

30

Corporations and the Community

For as long as I have been in the business world, there has been debate over whether corporations should be active in philanthropic activities. Some take the position that any funds earmarked for charities should be passed on to the shareholders, who can decide on their own where they want contributions to be sent. Warren Buffett, of Berkshire Hathaway Inc., is very much in this camp.

On the other side are people who argue that corporations exist and thrive as part of a society and therefore have a responsibility to support the health and vitality of that society.

Before I discuss my own views, let's put the subject into some perspective. Charitable contributions in the United States in 2001 totaled more than $203 billion (up from $85 billion in 1970). Corporate giving was only $11 billion of that total and represents about 1.2 percent of corporate pretax income. This is a percentage that has remained largely unchanged since 1970. Outside the United States there is very little tradition of corporate giving to nonprofit organizations.

I come out squarely as an advocate for the second point of view. I believe corporations succeed only if they operate in a healthy and vibrant society. They need the communities where their customers and employees live to be strong, as much as they need successful research, planning, and advertising. Contributing to their communities is, therefore, good business, too.

However, in one respect I am closer to the Warren Buffett view. I have little enthusiasm for what I call "checkbook philanthropy." Corporations that regularly allocate a certain amount of money from their budgets for philanthropic activities and then dole it out over the year to miscellaneous charitable organizations are certainly doing some good. However, I believe they are underperforming in a substantial way.

As the numbers I just quoted indicate, corporate giving is not a significant part of the cash contributions that go to charities in America every year. A surge in personal giving could make up for all of corporate philanthropy cash without a lot of effort.

My strongly held belief is that corporations can and should play a role far beyond simply writing checks. Corporations, as distinct organizations in our society, do certain things better than all the other parts of our society. Most important, they know how to plan, manage resources, communicate to constituencies, and conduct many other productive activities that are also required by nearly all nonprofit organizations. Skills in these areas are very important to charitable organizations, but such skills are rarely found in sufficient quantities to allow the emergence of successful, self-renewing organizations. Where else can these organizations turn for the help they need in building and maintaining organizational excellence? Certainly not governments, whose skills seem to be limited to giving away money and writing regulations. Individuals can be enormously generous with their financial resources and can provide priceless value through the hours they volunteer, yet they lack the heft and leverage that corporations can bring to the solution of large systemic problems.

Let me give you a very real example. My great passion in life for the last thirty-five years has been fixing the public school system in America. Beyond my family, my Church, and my work, this has been the center of my life. In 1995 I spoke to the National Governors Association (the organization of the state governors of the United States), urging them to step up their efforts on public school reform in their states. Their response was swift and unanimous. They said to me: "We agree with you, Lou, and want to do a lot more. However, we cannot do it without the help of the business community. We need the business community to energize and drive change in our state legislatures, in our school boards, in our school administration bureaucracies. We need you standing side by side with us explaining the urgency of the problem and demanding that tough decisions and changes be made."

As a result of that meeting, an organization called Achieve was created. It has, arguably, been the most important force for public school reform in America in the last decade. It is a combination of active governors and chief executive officers who have worked to build a consensus and a momentum behind standards-based education reform in the United States.

This is clearly not checkbook philanthropy. This is hard work. It's not glamorous. It doesn't make the front page of the newspapers, yet it is a task—as the governors said emphatically—that CEOs and their corporations can carry out uniquely for our children.

Making Real Change

If every corporation in America thought about how it could bring unique skills and resources to the solution of societal problems, the positive impact would be many, many multiples of the $11 billion in cash.

Rosabeth Moss Kanter, a professor at Harvard Business School, calls this a paradigm shift from "spare change to real change."[1] She suggests that companies can move from social responsibility to social innovation by viewing community needs as opportunities rather than simply seeing them as responsibilities. She points out that many recipients of corporate largesse don't need charity. They need change. I agree.

My first experience with corporations unleashing their power to help philanthropic causes was at American Express. In the early 1980s we introduced a concept called Cause-Related Marketing. We told all of our cardmembers that we would give 1 cent to the restoration of the Statue of Liberty for every charge on the American Express Card, plus 1 cent for each purchase of Traveler's Cheques, $1 for each new card enrollment, and $1 for every travel package worth $500 or more. The response was truly amazing. Nearly $2 million was raised in a very short period of time. However, while the money helped, it paled in comparison to the benefit that the Statue of Liberty Organization received in the form of savvy, sophisticated American Express marketing. We literally opened the nation to the organization's cause, blanketing every community with its message. This was a bona fide case of moving from short change to real change.

At IBM, a sense of—and commitment to—corporate responsibility has always run deep. In the decade before I arrived at IBM, the company's philanthropic contributions totaled $1.3 billion, making IBM one of the world's most generous companies. But virtually all of IBM's contributions were made in cash, and they were spread across a multitude of issues and geographies. Some grantees were highly dependent upon IBM's grants and had been IBM grantees for decades.

I felt instinctively that we could be more effective if we focused and targeted our efforts on solving problems and not merely throwing money at the problems. As we reengineered the rest of IBM, we reengineered our philanthropic philosophy as well, focusing on the

---

1   "From Spare Change to Real Change: The Social Sector as Beta Site for Business Innovation," by Rosabeth Moss Kanter, *Harvard Business Review,* May–June 1999.

use of technology as a way of solving social, and especially education, problems.

Essentially we began to treat our grantees as if they were customers and offered them access to the best technology and talent in the company. We helped them solve problems by rolling up our sleeves and getting into the trenches and partnering with them. We launched an initiative called Reinventing Education. It began in the IBM research laboratories, where teams of IBM researchers worked in tandem with teachers and administrators—and often students and parents—to eliminate the toughest barriers between our children and a world-class education: problems concerning how teachers are trained; or how time is used in the day, week, or year.

Our researchers used voice recognition to invent a new way to teach children to read, digital portfolios to evaluate teacher skills and student learning, data warehousing to inform decision making, and online programs to help teachers access the best lesson plans and learn how to correct student work. In 1995 we began a process of working in states, school districts, and in countries outside the United States to help our education partners learn how to use and adapt these technologies. The program now reaches 10 million children and 65,000 teachers and has increased student achievement across grades and subjects.

Most striking is the fact that the largest expansion of these programs and institutionalization has occurred after the IBM grants ran out, indicating that our school partners have taken these innovations and internalized them. This is as true in Vietnam and in Brazil as it is in West Virginia and North Carolina.

The concept of bringing our unique resources (technology and talent) was applied to areas beyond education, like the United Negro College Fund and the tragedy of September 11, 2001. Our employees have endorsed our strategy with personal commitments as well: In 2001 our United States employees contributed 4 million hours of their time and nearly $50 million of their own money to a range of social and educational organizations. Nearly 10,000 IBM employees serve on the boards of those organizations. We support these employees in several ways. We offer access to a fund that makes technology grants in small amounts to organizations in which IBM employees volunteer more than 100 hours a year. For the 5,000 IBM employees who serve as electronic mentors to students in grades K–12, we offer Web-based support that provides tools and special content. We also provide a 4–1 match on employee gifts to K–12 schools.

I hope more companies, small and large, will identify opportunities in which they can make a difference—in which they can infuse their special resources and talents, not just money, into our communities.

Approach the task like a business opportunity: Here is a need. What resources and programs are required? How do we organize to get the job done? How do we ensure that we are measuring results and not just activity?

In no other area of corporate endeavor—not in advertising, not in research, and certainly not in marketing or manufacturing— would we evaluate our effectiveness simply by measuring how much we spend. Why should philanthropy be unique?

[31]

IBM—a Farewell

As I take my leave of IBM, I am experiencing a flood of unexpected feelings.

I left McKinsey at the age of 35, happy with what I'd learned and raring to go to a new life as an operating executive. Partners frequently depart McKinsey to join their clients; I was simply following a path created by others.

I left American Express after eleven years primarily because I disagreed with the overall corporate strategy being pursued at the time. Had events been different at American Express, I might have remained there to this day.

As I have already said, the leveraged buyout of RJR Nabisco was doomed from the start, so my leaving there was a well-timed exit, but it was also a response to the enormously exciting challenge of trying to lead the turnaround of IBM.

My feelings as I leave IBM are quite different from those earlier transitions. With all my other moves, I was always looking forward to the future, to new challenges. While I am excited about the new life I am constructing for my post-IBM years, I find myself in the final months of my IBM career looking back—more than I have ever done before.

I came to IBM as an outsider, a force for change. I had to make a lot of difficult decisions, wrench the company in ways it did not want to go. Along the way, in my heart I became a "true Blue" IBMer. Interestingly, this outsider occupied the CEO's office longer than any IBM Chief Executive Officer other than the Watsons. Here is the letter I wrote to my colleagues to announce my retirement:

January 29, 2002

L. V. Gerstner, Jr.
Office of the Chairman and Chief Executive Officer

Subject: CEO Transition

Dear Colleague:

When I joined IBM on April 1, 1993, there was no thought about my retirement date. The Board of Directors asked me to focus on one short-term objective: save the company. Given my very limited knowledge of IBM at the time, I quite honestly did not know if that could be done. I certainly didn't know how long it might take.

Well, with the support and leadership of thousands of IBMers, we did turn the company around. That work, and my original mandate, was largely completed by the mid-nineties. But along the way, something happened—something that, quite frankly, surprised me. I fell in love with IBM. I decided, like many of you, that this was the best company in the world at which to spend my career. IBM is a fascinating, important, frustrating, exhausting, and fulfilling experience—and I've enjoyed every minute (well, maybe not *every* minute)!

But here we are nearly nine years later, and now it *is* time to address retirement. I have always used these e-mails to speak

with you about the most important developments and our strategic direction. I want to do that again now. Moments ago, the
Board of Directors elected Sam Palmisano to be Chief Executive
Officer of our company, effective March 1, 2002. Also, John
Thompson, Vice Chairman, announced his intention to retire
from the company and the board on September 1, 2002. I know
the entire IBM team joins me in thanking John for thirty-six stel-
lar years of IBM service—a wonderful career that included building our software business, and that culminated with focusing the
way we identify and pursue new market opportunities.

At the board's request, as well as Sam's, I will remain as
chairman until the end of this year. From March 1 on, Sam is our
new leader. My job will be to help him in whatever ways he seeks
my time and counsel.

Let me say something about the timing of this transition,
because some people believe IBM CEOs are required to step aside
at age sixty. That's not so. There is no rule or age limit that requires me to do this now. I am doing it because I am convinced
that the time is right. The company is ready, and so is the new
leader. I have never felt more optimistic and confident about our
future. And those are the best circumstances under which to
make this sort of change.

Over the past two years, Sam and I have forged a strong
partnership to prepare the company for a transition in leadership. Supported by a fine Board of Directors, we have undertaken
a process that has been disciplined, transparent, and thorough.

Many of you know Sam. Thousands of you have worked for
him. He's an exceptional leader, passionate about our business,
committed to our principles and values, and steeped in the dis-
ciplines that are critical to our success. Beyond those critical
qualities, Sam bleeds Blue. And because he does, he understands
the character of our company at its soul, the incredible world-
changing things it alone can accomplish—and how it must con-

tinue to change in the years ahead. I know you will give Sam all
the support you so generously provided me over many years.

It has been the privilege of a lifetime working with you
these past nine years. I am so proud of so many things that we
have accomplished, far too many to list in this e-mail. All our
hard work has brought IBM back. Today, our strategies are correct.
Our capacity to innovate is unmatched. Our culture is moving in
the right direction. And we have restored the pride all of us feel
in this company. Those were pretty remote targets back in 1993,
when so many had written us off and so few believed we had the
will to survive. But in your gritty, classy, determined way, you
never gave up. Thank you for restoring IBM's leadership.

As I said, after March 1, I'll be available to help Sam and
the entire leadership team in any way I can. And long after I step
aside as chairman, I want you to know that I will be cheering and
rooting for this magnificent company and its extraordinary people. I am an IBMer for life.

As much as I meant those words, and as much as I treasure the
thousands of wonderful expressions of appreciation I received from
IBM employees, I now realize that I was always—even to the end—an
outsider.

My most senior colleagues—Sam Palmisano, John Thompson,
Nick Donofrio, and others—who worked side by side with me and
deserve as much credit as I do for IBM's renaissance, share a perspective I will never have. They have lived their business careers at IBM.
They have seen it all: the glory days, the agony days, the turnaround
days. Their roots are deeper than mine, their experience richer.

For me, IBM pre-April 1, 1993, is a mansion full of many rooms
but without doors. I never entered it. I dreaded going into that house.

I had to drive change, and I knew that all the reasons not to change were in those rooms. I can recall numerous occasions in the early days when I would outline a change I thought was necessary, and my team would say: "Oh, we tried that before and it didn't work." I couldn't explore the "befores" or I'd learn all the reasons not to change.

Yet on occasion I would hear my colleagues reminisce about special experiences, both joyous and painful, that shaped their lives or shaped the company. I recently asked one of my colleagues to tell me about the CEOs who preceded me—the men who took over from Tom Watson. It was a fascinating hour for me, and I wish my time at IBM would have allowed me to make the connections from the old to the new; not the strategic or the cultural connections, because in a sense we did a lot of that. Our strategic moves had much to do with returning IBM to its roots as a research-driven builder of large systems and infrastructure. Our cultural transformation sought the high-performance culture that animated IBM under both Watsons.

The connections I could never make were the personal linkages—the laughs and the tears of joining a great company together, training and growing together, winning and losing together.

Yes, I was always an outsider. But that was my job. I know Sam Palmisano has an opportunity to make the connections to the past as I could never do. His challenge will be to make them without going backward; to know that the centrifugal forces that drove IBM to be inward-looking and self-absorbed still lie powerful in the company. Continuing to drive change while building on the best (and *only* the best) of the past is the ultimate description of the job of Chief Executive Officer, International Business Machines Corporation.

APPENDICES

The following material provides amplification of points made in the text of the book.

Appendix A: Examples of employee communications
Appendix B: The future of e-business
Appendix C: Financial overview of the IBM transformation

Appendix A

Included here are a number of memos (of the hundreds) that I sent to IBMers. They were selected to illustrate the range of subjects and the tone of voice that underscored communications as a vital step in our transformation. Most were sent to all employees. Some went only to managers or senior executives.

Crisis Letters

In a crisis it's far easier for the company to emerge intact if the CEO makes sure that all the employees know there is a crisis, what the management is doing about it, and what everyone must do to help. The following letters cover three major crises that IBM faced over the last ten years: 1) when we decided in 1993 to take more than $8 billion of cost out of the company; 2) when the possibility of Y2K computer problems had all of our customers on edge; and 3) when the tragic events of September 11, 2001, created deep employee anxiety and major customer problems.

July 27, 1993

Office of the Chairman

MEMORANDUM TO: All IBM Colleagues

SUBJECT: Today's Announcements

I want to take this opportunity to tell you where we are as a company, especially in the context of today's earnings and re-structuring announcements.

When I first wrote to you back in April, I said I arrived at a very painful time for IBM. Unfortunately, the pain is not behind us yet, as evidenced by today's events. As our second-quarter re-sults show, IBM is not profitable with our current cost structure.

I also said in April my first objective is to get our company right-sized as quickly as possible. That, along with getting as close as we can to our customers, remains one of my highest pri-orities.

Making reductions slice by slice, quarter by quarter, as we have over the past couple of years is unfair and debilitating. It creates anxiety for you and uncertainty among our customers. So we have got to get the company to the right size, and then focus all our energies on growth. Today's announcements, when fully implemented, are intended to do that.

I know you have heard this before, and each time you thought it was the last time. But if our current view of future rev-enue and demand is accurate, I believe we will have turned the corner.

Of course, there always will be a need to adjust our struc-ture to reflect market realities. But we expect to be able to do so in the future without these large adjustments.

I know these actions may be bigger and deeper, and may take longer, than many of you anticipated. They certainly have kept me awake at night. And I wish I could be specific about the details of how this will affect each of you.

I can tell you this much: All of our businesses have determined what overall workforce size and capacity levels will make them competitive. They are still working out how specific locations will be affected. I have asked the general managers to act as quickly as they can.

Wherever possible, normal attrition and voluntary programs will be used. But in some cases, layoffs may be required, and you will be informed as soon as we have answers.

I can assure you, these actions must be taken to restore IBM to profitability. We are not generating any profits to reinvest in the future of our business. And that's critical because without funds to invest—in technology, in people, in new markets—there is no future for any of us at IBM. Profitability brings stability; stability brings growth; and growth brings the security and well-being we want for all of us.

You and I want to make IBM the most successful company in the world. It was once, and I am convinced it will be again. In my first few months here that conviction has grown, day by day, customer by customer, and employee by employee.

Since the day I arrived, I have made it my business to talk to as many of you, and as many customers, as I possibly could. From what I learned, there is a lot more right than wrong with this company; and I'm convinced our weaknesses can be overcome by our strengths.

Our people are skilled and committed. Our technology is outstanding. Our customers want us to succeed. And our size and global reach give us a competitive advantage unmatched by anyone, anywhere.

We have a lot to be proud of, and much to be excited about

in IBM. Our future is bright because of our great strengths. But we need to get the problems behind us and start focusing on growing again. That's what today's programs are intended to do.

I thank you very much for all you've done in the past, and I ask for your patience and continued support as we move forward in the months ahead.

Lou Gerstner

December 28, 1999

L. V. Gerstner, Jr.
Office of the Chairman and Chief Executive Officer

Subject: Counting Down to Y2K

Dear Colleague:

As we head into the final hours leading up to Y2K, I want you to know where we stand.

While no one knows with certainty what will happen in every part of the world, we do know a few things. The vast majority of our customers—particularly those representing the world's largest businesses and government agencies—tell us they are ready. These include banks, telephone and power companies, hospitals, airlines, and air traffic control authorities. They have tested and retested their systems, and many have said publicly that they are prepared. We remain less certain about the readiness of developing nations and of small businesses. We know there will be outages and problems in places, but most of these should be identified and fixed fairly quickly.

We know one more thing about Y2K. We know that our company has expended incredible energy and effort to help our customers and to get our own internal systems ready. This has involved great sacrifice on the part of IBMers—and their families—everywhere. I know that tens of thousands of us have changed our holiday plans so we can pitch in during these final days. Every customer support representative in the company will be on the job or on call this weekend. Thousands of other IBMers have volunteered to be called on, if needed.

As Y2K hits—first in New Zealand, then moves across the

world, time zone after time zone—we can all take pride in the fact that IBMers will be at the ready. Internally, we're battened down. Externally, we're poised to help. I want to thank all of you for this unprecedented effort. While we all hope that Y2K turns out to be the biggest nonevent of this millennium (and the next), this kind of commitment to customer and team is IBM at its very best.

w3.ibm.com will post updates throughout this critical weekend. I'll be back to you with a report after January 1.

I would like to take this opportunity to wish you and your families a happy new year, an exciting new century, and a peaceful new millennium.

January 11, 2000

L. V. Gerstner, Jr.
Office of the Chairman and Chief Executive Officer

Subject: Y2K Perspective

Dear Colleague:

Now that we've got a full business week behind us, it's clear that the Y2K transition, both with our customers and within IBM, has gone far better than we hoped or expected. We saw a very small number of potentially serious problems last week, but we fixed all of them quickly. While we will remain vigilant in the weeks ahead, we can all feel very good about what's happened—or, rather, about what didn't happen.

While there was never any real concern about a "digital winter" enveloping the globe, the relative calm our company and our customers experienced could only have happened as the result of solid planning and hard work on the part of tens of thousands of IBM colleagues around the world.

On New Year's Eve I had the opportunity to spend time with the team in our Millennium Center in Southbury, Connecticut. It's hard to express the pride I felt talking with this team of professionals and seeing them in action, knowing that they represented just a fraction of the IBMers worldwide who had worked and sacrificed to see our company and customers through this event. I want once again to thank the colleagues who were at their stations during this critical period and all the others whose work over the last few years was responsible for this resounding success.

In reviewing the extent of this undertaking, I am awed at the effort IBM and IBMers put forth:

- More than a million items critical to our own operations were readied for the transition—from PCs and servers, to application software, manufacturing tools and instruments, 2,500 suppliers, 750 business partners, and 200 subsidiaries.
- More than a million customers used our technical support Web site.
- IBM Global Services processed more than a billion lines of customer code.
- During the crossover period, some 50,000 colleagues were on the job or on call.

Now, we're into a new year. The fact that Y2K went so smoothly for customers is great news for them—and for us. It may mean that customers who locked down their systems to prepare for the transition are going to unleash a lot of pent-up demand sooner rather than later. We must seize these opportunities and get the year off to a very fast start. I will have more to share with you about our near-term priorities after we announce fourth-quarter results next week.

September 11, 2001

L. V. Gerstner, Jr.
Office of the Chairman and Chief Executive Officer

Subject: Update—Today's Tragic Events

Dear Colleague:

Scores of IBM colleagues have reached out to me and the company's other senior leaders to ask how they can help in response to today's catastrophic events in the U.S. Both as individuals and as an institution, there's a great deal we can do—and are already doing.

As individuals, we can answer the calls that are going out for blood donations, particularly in the New York and Washington, D.C., areas. Of course, we can and should also reach out to colleagues who are distraught over today's events.

But perhaps the greatest contribution we can make is as a company. I don't need to tell you that the object of this kind of attack is to disrupt and paralyze. Already today, many national leaders have said that we can't allow even acts this heinous to shut down the operations of the United States. In the past few hours we've heard from major customers who will need our help to resume operations, and IBM teams are gearing up to respond. So, as we did with the work to prevent a Y2K "meltdown," we now have an important institutional role to play in the restoration of the infrastructures that have been devastated. The best thing we can do at this point is stay focused and remain ready to assist if called on.

In response to some specific issues, such as air travel, or making customer calls in New York City and Washington D.C., I'd urge you to take guidance from the authorities who are in the best position to make those decisions. And as I wrote to you earlier today, continue to exercise discretion and use your best judgment as you conduct business.

Finally, we are doing everything possible to locate and account for our colleagues in New York City and Washington, D.C. Our efforts will continue around the clock. We will keep you updated.

I want to thank you for the concern you've expressed and for the work many of you will do in the days ahead.

September 13, 2001

L. V. Gerstner, Jr.
Office of the Chairman and Chief Executive Officer

Subject: Update on Tuesday's Events

Dear Colleague:

I want to update you on where we are and what we've done since Tuesday's tragic events.

First, and most important, we have accounted for all but a handful of our IBM colleagues who might have been in New York City or Washington, D.C., when the terrorists struck. Of course, we will not stop until we have accounted for every one of our people. I know each of us is hoping and praying for a good outcome.

Sadly, as we have reported on w3 [IBM's employee Web site], we received confirmation on Wednesday that one of our colleagues was aboard one of the hijacked airliners. I know all of us are deeply grieved by this news. In addition, we have been learning of IBMers whose family members were killed or injured. Moments ago, I heard from an IBM colleague whose daughter was also on one of the hijacked planes.

Words fail to convey my sadness when I hear such devastating news, but on behalf of all IBMers worldwide, I wish to express our condolences to the families and friends of all those who have lost loved ones.

Let me update you on what we are doing to help customers. You may be surprised to learn that more than 1,200 IBM customers were located in the World Trade Center or within a two-block radius. Hundreds of them have contacted us since Tuesday morning. Currently, we're managing or have already resolved 20 full-blown emergency situations. We're rolling in large servers, thousands of ThinkPads, and workstations; we're providing thousands of square feet of data center capacity; re-creating data processing environments that were destroyed; and relocating customers' operations to IBM facilities. In addition, we are helping various disaster relief organizations with IBM products and assistance. Thousands of our colleagues are on the case, and the work proceeds around the clock.

I continue to receive hundreds of notes from IBMers all over the world. I trust you understand that I cannot respond to each of them, but I want you to know that I read every one. I have been deeply moved by the outpouring of concern and, most of all, your compassionate offers to help in any way possible.

There are plenty of opportunities for individuals to help. Those of you who have offered your time and skills may yet be called on, so stand by. Many have asked if we're going to run blood drives at IBM facilities. We have been in contact with the Red Cross and have been advised that the best way to provide blood is to donate it at the local community level. As it happens, several IBM locations in the U.S. were planning blood drives this week and next. These will proceed.

A number of relief funds have been established by government and volunteer agencies, and I know from your notes IBMers will be extraordinarily generous, as you have been in a number of prior national emergencies. We will provide on w3 information on ways individuals can contribute.

A special fund, called The September 11th Fund, has been established in New York City by various organizations, including the United Way. This fund will deliver financial services and assistance to those who were affected by Tuesday's catastrophe. IBM has pledged $5 million in cash, technology, and technical assistance to this fund. This is in addition to the uncountable product and human assistance IBM is providing to other agencies and organizations to help them manage through the crisis.

As I wrote to you on Tuesday, the most important thing any of us can do is take care of the job at hand and keep IBM moving forward. I ask you to remain focused on your customers, your job—wherever you are in the world—and trust that the local teams in New York and Washington, D.C., will reach out for all the additional assistance they need.

Your concern and self-sacrificing spirit make me so proud of our company and of each other. Let's stay focused, and stay together.

September 21, 2001

L. V. Gerstner, Jr.
Office of the Chairman and Chief Executive Officer

Subject: Our Response to the Crisis

Dear Colleague:

I want to bring you up to date on how our company and our people are responding to the traumatic events of September 11.

In my previous notes to you, I've written about the effort to account for more than 2,000 IBMers in the affected areas, including all those with home addresses near the World Trade Center. The effort was immediate and exhaustive. I want to thank all of our colleagues who helped in this vital work, especially our managers and team leaders.

As you saw on w3, one of our colleagues, Zandra Cooper Ploger, was aboard one of the hijacked jetliners. Zandra's coworkers will tell you, as they've told me, that she was a true professional and the kind of person you were simply glad to be around. As I write this, another colleague, Brian L. Jones, is still listed among the missing at the World Trade Center. Brian was last seen on the 78th floor, helping coworkers and his customers from Franklin Fiduciary evacuate the building. I speak for every IBMer in keeping hope alive for Brian, in mourning the loss of Zandra, and in extending our sympathies to all IBMers who lost family members or friends.

It's not possible to capture, in a single note, the scale of the response that was mobilized on behalf of our business and government customers, literally within minutes of the first attack. But behind every one of these customer stories, behind hundreds of servers and 5,000 workstations shipped, 12,000 ThinkPads configured, scores of applications brought back on-line, and more than 100,000 square feet of temporary facilities outfitted for customers, there are IBM people. The client teams mobilized, of course. Behind them, supporting them, have been 600 customer engineers and customer service representatives, plus 200 project managers and consultants in our business recovery unit alone—not to mention at least another 1,000 services people locally and thousands of colleagues across the company assisting in this work.

I saw it myself at our business recovery center in Sterling Forest, New York, on Tuesday. I watched as IBMers in the command center put calls out for help—and as their colleagues from across the company practically jumped through the phones to respond. Virtually every major business unit was on the scene, truly working as one team. It was IBM at its very best.

Every one of the customers I met at Sterling Forest asked me to tell you all how much they appreciated your efforts. And we've received call after call from some of the world's leading financial services institutions—from lower Manhattan and from some of the largest European and Asian banks—delivering their personal thanks for the superhuman effort they got from their IBM teams. One CEO told me: "IBM stands alone in their response to this crisis." On Monday of this week—when the New York Stock Exchange reopened to a day of record volume—the president of a major brokerage led his team in a round of applause for IBM.

I know you share my pride in our IBM colleagues who have performed so magnificently on behalf of our customers. We can be equally proud of the IBMers who led our response to the relief and rescue effort. We immediately supplied ThinkPads and workstations to the Red Cross and built a secure communications network for field workers at Ground Zero in lower Manhattan. With cellular service crippled, our mobile solutions team created a wireless e-mail network for handheld communications among workers who are locating the disabled and elderly, and expediting aid, medications, and evacuations. Another 200 desktops were shipped to family assistance centers in New York, and we provided a Notes/Domino solution to help the U.S. Federal Emergency Management Agency. Some 500 IBMers made all of this happen.

In addition to our $5 million contribution to The September 11th Fund—which we joined with the United Way and other organizations to create—IBM immediately took over management and hosting of its Web site. Already, the fund has received more than $100 million in pledges.

I want to spend just a moment on one other aspect of our response. This can't be quantified as easily as numbers of ThinkPads shipped, or dollars donated, or customers helped. But years from now, it will be my most enduring memory of how you rose to the challenge of September 11. It is all documented in the tidal wave of notes—thousands—I have received from IBMers looking for any way to help.

Typically, the notes would begin: "I'm a DB2 specialist in Atlanta . . ."; or "Although I am in London, I have extensive experience in UNIX . . ."; or "All of us in Region 10 . . ." People who couldn't get on planes offered to drive hundreds and, in some cases, thousands of miles. One colleague told me if he didn't have the skills required, he'd go out and get them.

I heard from colleagues in Dublin who were getting extra pay for working a public holiday. They wanted to donate it all. Others offered to donate their variable pay or vacation pay. An award-winning server sales specialist wanted to know how to "return" a seven-day vacation so that the funds could be contributed to aid the victims. One pSeries product manager won a new sports car in another sales contest. He asked that IBM instead donate the cash equivalent to the relief effort. His note to me said: "This is the least I can do for those suffering and searching for loved ones."

To all who volunteered, you have my sincere thanks. Some of you will be called. Many others will be called upon in a different way: to keep our great company moving ahead.

Easily half the notes I've received have come from outside the United States. Of course, the September 11 attacks victimized not just American citizens and American companies, but citizens and institutions from across the world. I've seen reports that people from 80 countries were lost. But I want our colleagues outside of the u.s. to know how especially heartening I found your notes of sympathy and solidarity. This *has* been a profound disturbance in the United States. This is a young nation. It has rarely experienced, at home, the kind of savage attacks that sadly are part of the history and current reality of many other parts of the world.

There's nothing pleasant about having to confront a situation like this one. But since all this started last Tuesday, the great strengths of IBM and its people have been borne out again and again, in countless acts of professionalism, and caring, and personal sacrifice.

I've seen our strength also in the notes some of you have sent me—expressions of outrage at the mindless hostility shown to people of Middle Eastern heritage, and words of pride and gratitude that in a global company like IBM such behavior has never been, and never will be, tolerated.

With time, all the lessons of September 11 will become clear. But already we know this: Ours is a strong and compassionate company. And tough situations like this bring out our true character, every time. We've seen that over the last two weeks, just as surely as we saw it in the millions of dollars and thousands of hours IBM poured into the recovery from disasters like the 1995 earthquake in Kobe, Japan, and many others over the years.

Long before I came to IBM, I knew this was a company of great integrity and character. I can see now that I couldn't fully appreciate just how deep that ran, or how strong and good the people of this company are. When the need was the greatest— when our customers, colleagues, and communities needed it most—you stood tall. I have never been prouder to call myself your colleague.

Culture Change Letters

One of the most frequent themes of my letters to IBM rank-and-file employees—and also to managers and executives—concerned the changes we needed to make in our culture, and why we needed to make those changes. I frequently picked mundane announcements, like organization changes, and turned them into opportunities to underscore important culture themes.

July 6, 1994

L. V. Gerstner, Jr.
Office of the Chairman and Chief Executive Officer

To: Senior Leaders

Of all the issues I have wrestled with since joining IBM, the subject of corporate officerships has been the most perplexing. I've seen many other companies during my years as an executive and consultant, and I've found no company with a system comparable to IBM's officer program. To make sure, I had this confirmed through an extensive survey that was recently completed.

Nearly all other companies use corporate officer titles to identify people who occupy specific, senior corporate positions. At IBM, however, officer titles often do not stay with positions, but are awarded to individuals and then follow that person from job to job. This creates a bewildering array of problems, including these:

- two or more people holding the same job and being accountable for the same kinds of results, but some are officers and some are not

- incumbents entering a job previously held by an officer and not being accorded that title
- individuals enjoying the perks associated with officers, yet not performing at the level of others who carry more significant responsibility
- executives turning down important new opportunities and responsibilities because they fear internal perceptions—that the new job is not prestigious enough for a corporate officer title
- on at least one occasion, officers reporting to non-officers

Most significantly, I believe, this program is viewed as mysterious and frequently subjective. As a result, it accentuates our tendency to be preoccupied with internal affairs as opposed to external successes.

I have always believed that titles should be attached to positions and not individual people, and that when we make a decision that a person is capable of filling a given job, he or she should carry the same title as others who had that job. I also believe that rewards should be based on performance in the marketplace, teamwork, and leadership, and that all of us—myself included—should be appraised on a regular basis and not tenured because of some earlier recognition.

Therefore, effective immediately, I am terminating the current corporate officership program. Corporate titles will be used only where appropriate for people who head significant corporate staff functions or have major corporate responsibilities (e.g., the corporate executive committee). Executives who do not have direct corporate responsibilities will have divisional titles. Like their corporate counterparts, business unit titles and in particular business unit officer positions should reflect function and responsibility. Those who currently have corporate officer titles, but who are business unit executives, will have a business unit title and will no longer have a corporate title.

Corporate Human Resources has been working with the business divisions on guidelines to provide title consistency throughout IBM.

A small number of you may feel that we are abandoning something of value—chipping away another piece of the proud old IBM. But our world has changed, and as we try to build a proud new IBM, this is one of many changes I believe we have to make.

My goal is to provide every person in this company with a clear career-development path based on performance and continuous achievement. I want all titles in this company to reflect responsibility and nothing more. I want all of our people to take on new assignments and new opportunities because of the challenges, the development potential for their careers, and the benefit that the company will derive from their performance. The future of IBM is in its human resources—our performers—and not in appearances and titles.

I cannot emphasize enough that in the IBM we are all working so hard to build, what is important is the quality of your work, the spirit of your cooperation and team effort, and your ability to contribute to the success and profitability of the entire enterprise. People who do good work—wherever they are, whatever they do, whatever their title—will be valued assets to the corporation, and their achievements will be visible to all.

Lou Gerstner

August 14, 1997

L. V. Gerstner, Jr.
Office of the Chairman and Chief Executive Officer

Subject: Changes to IBM Stock Ownership Guidelines

Dear Senior Management Colleague:

As members of the Senior Management Group, we are leading IBM into the future. Stock ownership plays a critical role in ensuring that our interests are closely aligned with those of our shareholders and in demonstrating our commitment to IBM. I firmly believe that a senior manager who owns a significant amount of stock in the company has a powerful incentive to maximize shareholder value.

IBM's stock has done very well over the past few years, reflecting IBM's strong performance and the movement of the stock market generally. Because of that—and because we have shifted to a total compensation strategy—we have decided to increase stock ownership guidelines to a multiple of total cash compensation, which includes salary plus incentive at 100% of target.

In line with the guidelines, I expect you to build investments in IBM stock through a combination of the opportunities provided. As members of the Senior Management Group, more opportunities are available to use for building ownership—through holding shares received from both the exercise of stock options and Long-Term Performance Incentive awards. We, as leaders of the management team, should be overweighted in IBM equity, not aiming for the same degree of diversification as an outside investor. That said, no one should feel uncomfortable about selling shares from time to time in order to accommodate occasional needs for extra liquidity.

We should be excited about IBM's future and demonstrate it in the best way we can!

July 10, 1998

L. V. Gerstner, Jr.
Office of the Chairman and Chief Executive Officer

Subject: Summer Jam

Dear Colleague:

Earlier this week I helped kick off a first-ever event hosted by IBM Research. Called "Summer Jam '98," it is bringing together 700 science and engineering summer students from eight different IBM laboratories to think about how technology will change the world in 2020. As in a musical "jam" session, they improvise, build on each other's ideas, and come up with provocative and exciting concepts, as well as some off-the-wall ideas. Here are a few of the early ideas I heard about:

- wearable computing technology like earrings and eyeglasses embedded with instantaneous language translation, speech recognition, and speech synthesis capability. Someone speaks to you in French. You hear it in English (or Russian or Japanese or . . .)
- a grocery store check-out machine that "isn't." Simply by picking up an item and taking it from the store, your noncash account is activated and the transaction is completed using "e-currency."
- "e-drink," a product that detects, through biometrics, who you are and, with "factory in a bottle" nanotechnology, instantly manufactures the beverage of your preference a moment before you drink it.

I certainly hope Summer Jam created at least a few seeds of ideas that really will transform the world (and IBM). For me, it was also a good reminder that:

- What matters in the marketplace—in the world—are solutions. Without ever consciously using that word, the students focused not on faster, smaller technologies, but on how those technologies can transform commerce, health care, family life, education, government, and human interaction. Our customers look to IBM to help them apply technology, and our ability to do that— drawing on all the technology and people resources of IBM—is our fundamental competitive advantage. We should never lose sight of that.
- We execute best when we work as a team. I continue to be inspired by what happens when a group of diverse IBMers combine their talent and knowledge to tackle an issue or an opportunity. The out-of-box ideas that are generated, the passions that are fired, are far more intense and interesting than what can be accomplished alone.

Summer Jam was a great event, but I don't think we need special days to remind ourselves that we work for a very important company in the most important and relevant industry in the world today. As our colleagues envisioned, information technology truly is changing the world in very meaningful ways. That's exciting for all of us.

July 10, 1998

L. V. Gerstner, Jr.
Office of the Chairman and Chief Executive Officer

Subject: Leading IBM

Dear Management Colleague:

I'm writing to you because you're one of the IBMers we've asked to manage people or lead a team. We've reached an important point in our mission to return IBM to industry leadership. The next step for us isn't dependent on a few senior executives making sweeping decisions like reducing our cost structure, restructuring our sales channels, or building our e-business strategy. Our strategies, for the most part, are in place. Taking the next step comes down to our ability to execute the strategies, and that depends on mobilizing all 270,000 IBM employees. I can't do that alone. Together, we are the leadership of IBM, and I need your help to move IBM forward.

I'm starting this dialog with you today because of something very interesting that happened earlier this week. As I said in the e-mail I sent to all employees today, I attended an event at IBM Research that was different from anything I've seen in my five years with IBM. It was called Summer Jam '98, and it gave me the chance to interact with a lot of very talented university students who are working here in technical positions this summer. I talked with them and took a lot of interesting questions.

One question stood out. The young man phrased it politely, but what he was really asking was: "I have a lot of good ideas that I'd like to see implemented in my lifetime. Why should I believe that will happen in a big company like IBM versus a small company or start-up?"

That's an important question. It gets to one of the most vital issues any leader in IBM faces today: why talented people sometimes leave our company, or why people we want to hire turn us down.

It's been my experience (and we have data that confirms it) that people don't make those decisions based solely on money or irrelevant things like their title. Most often people leave or hire on somewhere else because they feel their ideas aren't being heard, the business moves too slowly, we're focused on the wrong things, the bureaucracy is stultifying—or they're afraid it will be. In other words, it's about our "culture."

Who creates IBM culture? I suppose we could say it was created a long time ago. We could take the view that we inherited it, and passively accept it as "the way things are." The other option is to take some action. To lead.

Who accepts or rejects an IBMer's ideas? Who sets the pace, the tone for our team? Who determines priorities, makes investments, and recognizes results? Who encourages people to celebrate success, enjoy the wins, and learn from the mistakes? All of us who lead people create that environment. It's in our hands.

I think this comes down to a couple of things.

First, each of us must take seriously our role as a leader— the responsibility we carry to motivate and inspire the people around us. This is not a "nice to have" competency. Without leaders creating this culture, we can't recruit or retain the kind of talent we need. Far more important, we can't get superior results from the 270,000 people who are here. In the past, it may have been sufficient for managers to deliver the numbers and close the deal. Today, the definition of leadership at IBM is broader than that. You lead programs and projects, of course. But you're also in the job to lead people, to build a team, coach, and create a culture of high performance.

Second, you have some powerful tools and programs to help you do this job. We've invested $500 million in salary increases and $1.3 billion in variable pay this year. We nearly doubled the number of employees getting stock options in 1996, doubled it in 1997, and will more than triple it this year. These aren't "HR programs" that managers have to administer like some chore or annual ritual. They're powerful levers at our disposal to reward performance and encourage the behaviors and attitudes we need to succeed. I would say that managers today have more discretion and more latitude—and consequently have to exercise more judgment—in using things like compensation to reinforce performance than at any time in IBM history.

I hope I can count on you to exercise personal leadership—with imagination and passion for what you do, for our company, and especially for the people on your team. I can think of nothing more critical to our success.

We'll talk again.

September 28, 1998

L. V. Gerstner, Jr.
Office of the Chairman and Chief Executive Officer

Subject: Managing the Matrix

Dear Management Colleague:

I received a lot of feedback to the e-mail I sent to all IBM managers and team leaders in July. I was impressed with the serious level of engagement in your responses to that note, which talked about how we must create a work environment—a culture—where IBMers can thrive.

Today I want to talk about an issue that comes up virtually every time I meet with employees. It came up again two weeks ago during a meeting with some colleagues in Beijing. The issue, in essence, is: How do we deal with the IBM matrix? How can we execute our strategies in such a complex company?

Let me start by saying we can't hide the fact that IBM is a complex organization, and always will be. There are no enterprises that are both big and simple. They don't exist. So like any large organization, we have a management system that exists, in part, to deal with complexity.

More important, we must remember that IBM's unique value to customers is our ability to build integrated solutions, to be a trusted advisor who spans the full range of the industry. This is what sets IBM apart. It is our strongest competitive advantage. I'll go further than that. I believe our ability to integrate products, skills, and insight is our *only* sustainable competitive advantage.

We know there is no long-term, sustainable competitive

advantage in technology. At any given time, some competitor's nose will be in front of the pack with a faster server, a more powerful database, or networking product. I'm not minimizing the importance of technology leadership. We must fight and invest to stay ahead. But we cannot maintain, year-in, year-out, a marketplace advantage on technology alone. No company can. However, when it comes to integration, we've got a big lead on everyone else, a lead we should be able to sustain, and even lengthen.

The question then is: What kind of management system supports our fundamental strategy of integration? Many companies still have a traditional hierarchical system. It's relatively clean and clear. You know where decisions will be made, and who will make them. But the classic chain-of-command system is not only too slow for the pace of this industry, it opposes our ability to work *across* organizations. Hierarchy, which seems to support integration, actually fights it. Hierarchy erects vertical lines and boundaries and fosters the infamous "silo" mentality.

Now, there's no doubt a matrixed environment makes our jobs as managers and leaders much more challenging. It also makes them more important. While a badly managed matrix can be even worse than a hierarchy—in being both rigid and confusing—a well-managed matrix is highly fluid and adaptable. Roles change often. Teams form and disband. Decisions about which business unit will lead in any particular situation are not codified. This puts a premium on the judgment of leaders at every level. *You* decide which opportunities offer the highest potential return. *You* decide where to deploy our people, capital, and time.

Managing in a matrix is hard work, but it *does* work. Look at our recent multibillion-dollar win with Cable and Wireless in the UK. People and teams from every continent, multiple industry units, staffs, product and development groups pulled that deal together. Look at products like the Aptiva E and S Series, which feature technologies from across IBM. Look at the remarkable collaboration between Research and Storage Systems to set the bar higher and higher in disk drives. Or the teaming that goes on with our colleagues in Software Group, Lotus, Tivoli, and the sales units. Together, they've built one of our strongest growth businesses.

Unfortunately, we all know of many more examples where we've failed to work as a team, people focusing instead on their own priorities and targets, not willing to give—or accept—help or ideas. And, at the other extreme, people inserting themselves into *every* process, insisting on reviews and approvals. And meetings, endless meetings. All of this saps energy. It grinds the edges off our people.

I can only reach one conclusion. If something is a strength in one situation and a weakness in another, then there's not an endemic problem. We don't need to fix the "system" (though we'll continue to improve things like cross-company processes and the measurement system). What we need are leaders with the right attitude.

For example, if we have staff people who think their job is to monitor line people and turn them in when they do something wrong, no management system is going to succeed. But if our staff people have an attitude that says "My job is to ensure that all the capability of this company is channeled to the person at the point of contact with the customer. I'm here to help them win business," then the matrix becomes an asset because we're using it to deliver IBM value to the marketplace.

The good news is that every week I see more and more leaders who understand all of this, and live it. When they look at our matrix, they don't see a headache-inducing crisscross of reporting lines. They see a company flush with assets. They see a portfolio of people and products that our customers crave, and that our competitors envy and are trying desperately to replicate.

These leaders have another critical attribute. They trust their colleagues. They're not compelled to go to every meeting just to protect their interests. They say, "You've got the ball this time. What can I do to help you win?" And they understand that sometimes the best thing to do is get out of the way.

I trust I can count on you, as a member of IBM's leadership team, to exhibit the qualities that turn the breadth of IBM into a competitive strength, and to encourage that kind of behavior in your people. Remember: When you get down to it, the "matrix" is really Team IBM. Embrace it, don't fight it.

A final note. Many of you write to me on the leadership challenges we all face. While I can't answer every e-mail I get, know that I read every one with great interest.

August 23, 2000

L. V. Gerstner, Jr.
Office of the Chairman and Chief Executive Officer

To: Worldwide Management Council

Subject: On Committees

In response to several queries, I thought I would comment further on the committees that were announced as part of the recent reorganization.

First, a little bit of LVG management philosophy: Committees can become dangerous organisms. They should never be used as decision-making bodies. Their main purpose is communication—up, down, and across. They should meet infrequently with focused agendas, and they should be disbanded regularly to protect against the view that committees play important roles in successful institutions. They do not. Personal leadership and task-specific teamwork drive our success.

That being said, we all know it is absolutely essential that IBM work across organizational lines. Our unique competitive advantage is our ability to integrate within the company and in front of the customer. Therefore, there is a need for formal mechanisms for collaboration and communications.

Let me comment on each of the new committees:

Operations Committee

Sam Palmisano has been named Chief Operating Officer of the company. As such, I look to him to manage day-to-day operations and ensure that we are meeting our short- to intermediate-term targets. As we all know, meeting those targets in IBM requires close, friction-free interaction across our operating units. Sam will use the Operations Committee as a vehicle for communication—from him, to him, and among the team to ensure that we are working effectively as a team every day in the marketplace. Additionally, there will be a number of operational issues that will benefit from discussion among the members of the Operations Committee. These would include changes in our go-to-market model, planning for key customer meetings, pricing issues, etc.

Corporate Development Committee

John Thompson is now in charge of ensuring that our strategic visions are being converted to real, competitively assessed, market-driven, adequately financed plans and programs of execution. John will do this in many ways, but, for the most part, he will interact directly with the operating units. However, there will be times when an issue will be of such importance, or when John wishes to seek the views of others, that these strategy matters will be brought before the Corporate Development Committee. Most important on such issues, I look to the committee to enhance the quality of our thinking.

Additionally, the Corporate Development Committee will be the place where our corporate management team will explore our acquisition and strategic investment alternatives.

While individuals have been assigned to both of these committees, additional participants will be included, as appropriate, for specific subject matters. If the committees do not function as planned or if they need modification, we will change them with no hesitation. No compensation or perk distinctions will be associated with membership on these committees.

The Chairman's Council

This group, representing my direct reports, constitutes the corporate senior management team. It will not meet frequently but will be asked to advise me on matters of broad institutional importance, such as diversification, human resource policies, executive leadership and industry positioning, etc.

I hope this clarifies the important but limited role of these new committees. Please keep in mind that at least three other committees were eliminated as these new ones emerged. I suspect we will continue to modify these communication vehicles as we go forward. What will not change is our preoccupation with winning in the marketplace.

Making Employees Partners in Strategy and Implementation

I found early in my career that when I made employees aware of what the company was doing and why, everything became easier. In my IBM years, we had to make enormous strategic bets. I made sure our employees knew about them, and I gave frequent progress reports.

March 24, 1994

L. V. Gerstner, Jr.
Office of the Chairman and Chief Executive Officer

Dear Colleague:

As many of you know, shortly after I joined IBM last spring, I said that developing tough-minded strategies for each of our businesses was one of my highest priorities. Since then, we have moved aggressively on dozens of strategic fronts. We've talked candidly with our customers, from large- and mid-sized businesses to small-store and home PC users. We've taken long, hard looks at everything we do, inside and outside IBM, and made blunt assessments of what we do well and what we do poorly. We sized up our competitors. We looked at, and debated about, every product, every process, every piece of research and technology in our labs and in the pipeline.

We're not done yet. (In fact, we won't ever finish.) Nevertheless, we have under way today some 15 to 20 very specific strategic projects that will determine our future. They include task forces and implementation teams on things like PowerPC and Workplace operating system, software strategy, brand management, process reengineering, and alternative distribution.

For competitive reasons, we can't discuss the details of each of these efforts with the outside world. But I do want to give our shareholders some sense of our strategic directions. Consequently, I have aggregated and synthesized these strategic projects into six broad strategic imperatives. Call them a road map for IBM in the mid-'90s—where we need to go if we are to succeed in the marketplace, to grow, and to increase shareholder value.

This afternoon I will meet with securities analysts and with the media to outline these six directions. You'll be reading about these meetings in the papers tomorrow, but I wanted you to hear about them from me first. At the end of this letter are instructions on how to access a copy of my remarks, which will be available following my presentation to the securities analysts.

I don't believe IBM was lost in the woods the past few years. We knew that our industry was changing rapidly. We knew we had to change if we were to remain the industry leader—change directions and, more important, change how we work both among ourselves and with our customers. But perhaps prisoners of our own traditions, we were unable to do what we knew we had to. Simply put, we didn't execute.

Today, as in the recent past, execution—determined, quick, sure-footed implementation—remains our biggest challenge. In my presentation to the securities analysts, you'll see where we need to go. Our job now, every one of us, is to get us there.

I'm increasingly confident we can do it. In many parts of the company today, I notice a tighter focus, a growing sense of urgency, and a faster pace.

We have an exciting future. Let's step up the pace and get the job done.

Lou Gerstner

August 11, 1994

L. V. Gerstner, Jr.
Office of the Chairman and Chief Executive Officer

Dear Colleague:

Many of you in the Northern Hemisphere will be heading off to vacation shortly, so I wanted to take this opportunity to put some perspective on the first half of the fiscal year and to talk about what we have to do the remainder of the year.

We recently reported our third consecutive profitable quarter. Indeed, there is much to be proud of in our performance so far this year. We introduced several important new products, including mainframe, AS/400 and RISC System/6000 products, and ATM networks. We announced IBM Global Net and our new network services division. In both quarters this year our revenues increased, our profit margins remained stable, and we made significant, necessary expense reductions.

That's progress by any measure, and it's fair to say we seem to be gaining momentum.

But when you think about how difficult things were for us just two years ago, there is also a temptation to sit back and start humming "Happy Days Are Here Again"—to feel that things are back to normal once more and we can go back to being the good old IBM we all used to know and love.

We can't. Those old days are gone forever. Our industry continues to change at a breathless pace. Our competitors are continuing to make inroads into our market share. Our customers are reevaluating their whole approach to information technology. Industry profit margins continue to drop.

Rather than celebrate, we have to commit to focus intensely on two overall tasks: eliminating unnecessary cost, duplication, and bureaucracy; and implementing our business strategies.

As you know, our fourth principle says we will have a never-ending focus on productivity. Recent studies comparing IBM with our competitors show that we are still behind them in our cost structure. Many of our fundamental business activities—like inventory management, customer fulfillment, and information technology—remain uncompetitive. Also, as I travel around the company, I find we are still preoccupied with process, meetings, and bureaucracy.

We've got to fix all of this, and fix it fast. We have a major effort under way—organized under the general title of "reengineering"—to streamline decision making and our responsiveness to the marketplace. We've made some significant organization changes. But far, far more is needed in the months ahead.

Just as important, we have to implement the business strategies we have developed—specific results we must accomplish in all our businesses if we are to succeed amid all this change and once again be the most important force in our industry. I have a lot of confidence that we have the right strategies.

However, we've had outstanding business strategies before. I've read them all, and they were remarkably ahead of their time. The problem was, we never fully implemented them. We sat in meetings, nodded our heads in agreement, and then went back to doing whatever it was we were doing before. So we agreed we needed to change, but we didn't change. We said we needed new strategies, and we created them, but we didn't implement them. We said we wanted IBM to remain the leader in our industry, but we didn't do what we had to do to retain leadership.

What happened to us two and three years ago—all our

problems—was the result of this failure to do what we knew we had to do.

One of the most frequent messages you send me in e-mail is this: "I agree with everything you're saying about the need to change, but I don't see it happening where I work." Many of you are frustrated, and I'll tell you, I am, too, at times. I may be the chairman of IBM, but believe it or not, I run into pushback as you do.

Well, however strong the forces of inertia, we—a group of us—are going to make the changes we have to make. We are going to implement our business strategies. We are going to reenergize IBM.

We're going to start with our attitude. As you saw on my videotape speech last week, we need to make three commitments:

- a commitment to win in the marketplace
- a commitment to change
- a commitment to one another

No more focusing on our opponents down the hall. We're going to focus on our competitors, who have been focused on us for years.

No more endless meetings about the need to change, and then going back to business as usual.

I'm sorry, but making these commitments is not an elective process. This is the price of entry into the new IBM. We don't have room for pushback. We don't have room for spectators.

For those of you who have played an active role in making constructive change happen, I thank you and encourage you to intensify your efforts. You were largely responsible for our success the first half of the year.

For those of you who haven't yet made the three commitments, the train is pulling out of the station. Time to hop aboard or be left behind. I think the journey will be exciting and worthwhile.

Lou Gerstner

October 20, 1994

L. V. Gerstner, Jr.
Office of the Chairman and Chief Executive Officer

Dear Colleague:

Things are looking up.

A few minutes ago, we released our financial results for the third quarter. And, in case you haven't read the announcement yet, let me share the good news with you. IBM is profitable for the fourth consecutive quarter.

By any measure, we are making progress. We are sustaining the momentum we started to gain last quarter. In fact, we are starting to win in the marketplace.

The numbers are only one part of the story. The more interesting story is beyond what is captured on the financial statements. It's called execution of strategy.

There is growing evidence that we are starting to reap the benefits of our strategies. By working as a team—across divisional boundaries—we are beginning to leverage our size and scale. I see signs that suggest we are exploiting our technology better than we did in the past. Our efforts to increase our share of the client/server market are heading in the right direction and we are taking advantage of opportunities in key emerging markets.

Best of all, we're starting to deliver more value to our customers.

That's pretty impressive for a company that has been characterized by some as a dinosaur. Let me assure you, our customers around the globe are taking notice of our performance.

Eighteen months ago, we began a process to stabilize our financial position and to make our cost structure more competitive. Today our balance sheet is much stronger. While our overall costs are still too high, we are more competitive now than a year ago and in a healthier financial position. We can concentrate more of our efforts on serving our customers and growing our businesses.

All of this is not to say we've cleared the major hurdles. We still face big challenges. Our industry is full of opportunists ready to prey on any area of vulnerability. We can't declare victory or become overly optimistic.

To those of you who are playing an active role in our transformation—thank you. I recognize and appreciate your hard work.

Continue to persevere. Encourage your colleagues who are sitting on the sidelines to join our team. We have a formidable challenge ahead—one that requires the full participation of all IBMers.

Lou Gerstner

November 28, 2000

L. V. Gerstner, Jr.
Office of the Chairman and Chief Executive Officer

Subject: Chief Privacy Officer

Dear Colleague:

I want to tell you about an important new leadership position we are creating. It is important because it involves one of the most critical issues on the minds of customers worldwide—privacy in a networked world.

We know that one of the great conundrums of e-business is that it gives enterprises a powerful new capability to capture and analyze massive amounts of information—such as financial data and buying preferences and patterns—so they can serve individual customers more effectively. Yet this very capability troubles some people, who see it as a means to disclose or exploit their personal information. These are legitimate and very real concerns, and they must be addressed if the world of e-business is to reach its full potential.

At its core, privacy is not a technology issue. It is a policy issue. And the policy framework that's needed here must involve the information technology industry, the private sector in general, and—despite what some in our industry believe—public officials. Getting a constructive dialog going among all these parties is not particularly easy, but in recent years our company has stepped forward to do that. We have also worked hard on our own systems and practices to be responsive to customer concerns. We've made some progress, but more needs to be done.

That's why we're creating the position of Chief Privacy Officer. I am appointing Harriet Pearson to this new post. Harriet is the leader on our public policy team who has driven our privacy initiatives. She has been a passionate and expert voice within our company, our industry, and the general business community on privacy. Harriet will report to Senior Vice President Larry Ricciardi.

Externally, Harriet will represent IBM on this important issue. Internally, she will drive privacy policy across the company and serve as a resource for every business unit. She will coordinate a worldwide team of IBMers to unify all the work under way within the company—in our research and technology units, in marketing, sales, ibm.com, and our strategy and policy organizations.

IBM has a longstanding track record on privacy protection. I know this announcement will be seen by the marketplace as another example of our commitment, and that through our policies, practices, and technologies we stand accountable on this issue.

Thanks and Recognition Letters

As important as it was to tell everyone what big changes I had to make and why, it was just as important to thank people and to recognize the good work and the hard work they had done. We needed to focus on critical things, but we also needed to celebrate victories and heroes.

March 15, 1994

L. V. Gerstner, Jr.
Office of the Chairman and Chief Executive Officer

Subject: IBM Fellow

Dear Colleague:

As you know, an appointment as an IBM Fellow is the highest honor the company confers upon our scientific and technical employees. It recognizes sustained levels of distinguished achievement over a career and provides wide latitude to pursue technical projects of the IBM Fellow's choosing.

We normally announce new IBM Fellows at our annual Corporate Technical Recognition Event in June. This past weekend we bent the rules for a special reason.

Dr. Celia Yeack-Scranton of the Storage Systems Division was to be named an IBM Fellow in June. She was widely regarded as one of the industry's leaders in advanced magnetic recording. As an inventor and hands-on technologist, she made many important contributions to IBM storage technology. She was also a true team player, known for her ability to lead multidisciplinary teams.

Celia had a terminal illness, so on Saturday morning her manager visited her and told her she was named an IBM Fellow. She died yesterday.

We join her colleagues in Storage Systems and the Research Division in mourning her.

Lou Gerstner

May 12, 1997

L. V. Gerstner, Jr.
Office of the Chairman and Chief Executive Officer

Subject: Deep Blue

Dear Colleague:

I know I speak for IBM colleagues everywhere in congratu-lating the Deep Blue team on its outstanding performance. It was the culmination of years of research and exploration, and it will stand as a great example of IBM's technology leadership.

As much as I love to win (and I'm glad we did), I don't think the triumph of the match was that Deep Blue won and Garry Kasparov lost. The achievement was in demonstrating that powerful computers like Deep Blue can successfully tackle tough problems that require mind-bending high-speed analysis. Now we can apply what we've learned to help improve medical re-search, air traffic management, financial market analysis, and many other fields our customers care about.

I also want to thank Garry Kasparov. There aren't many peo-ple in the world who would have been willing to match their in-telligence and wits against an opponent like Deep Blue—and under intense media scrutiny. Mr. Kasparov never considered this match a sideshow. He took it seriously, and his sincerity as our partner in this experiment made it the invaluable learning expe-rience it was.

Lou

May 13, 1997

L. V. Gerstner, Jr.
Office of the Chairman and Chief Executive Officer

Subject: Stock Milestone

Dear Colleague:

Today we reached a major milestone in the history of our company. Our stock price hit 177⅛ during trading this morning, passing our all-time intraday high of 176⅛ on August 20, 1987. This is obviously good news for our shareholders—and, importantly, more than half of all IBM employees are shareholders, too.

But records are meant to be broken and milestones to be surpassed. In fact, I believe this is an early milestone in the long, long life of the new IBM. And other milestones are probably as important—achieving customer satisfaction second to none, accelerating revenue growth (we've posted record revenues two straight years, but we can do better), reaffirming technology leadership.

The next milestone I am looking forward to very much is just ahead when the total number of IBM employees will be larger than when we began building the new IBM in 1993.

In my eyes you stand tall. You did all this—the milestones passed, the victories just ahead, and those far down the road.

Thank you. Take a bow. You've earned it.

And, of course, I can't resist: Let's all get right back to work because we've just begun!

Lou

Communicating News

Employees in any institution frequently complain, and rightly so: "Why do I have to hear news about my company on TV or the radio before I hear it from the company?" Of course, we are required to break material news first to our shareholders through the media. However, I have always made it a point to make sure our employees hear the news literally seconds later. What follows is one example— when we announced that we planned to acquire Lotus Development Corporation.

June 5, 1995

L. V. Gerstner, Jr.
Office of the Chairman and Chief Executive Officer

Dear Colleague:

In keeping with my promise to share all items of importance with you, I want you to know that IBM announced plans moments ago to make a tender offer for 100% of the common stock of Lotus Development Corp., a leading software developer with whom we've had a long relationship.

At 1:30 P.M. EDT, we'll hold a press conference, which will be broadcast internally. But before I meet with the media, I'd like to tell you why we are making this move; why we think it's a win-win situation for IBM, Lotus, and the customers we both serve; and what's likely to happen next.

In calling this a win-win situation, I choose my words carefully.

IBM's strengths include a worldwide team of talented people, an unmatched customer base, a 140-country sales and distribution network, vast technical resources, a strong balance sheet, one of the world's most respected brands, and decades of experience in industrial-strength enterprise computing.

Lotus has an array of complementary strengths, the most important being its Notes and mail products that are rapidly gaining marketplace acceptance, a strong applications portfolio, a solid brand identity, and, of course, some of the industry's most innovative people.

Our goal, working hand in hand with Lotus, is to accelerate the creation of a truly open, scalable collaborative computing environment so people can work and communicate across enterprises and across corporate and national borders—without worrying about things like incompatible hardware and software.

People all over the world are seeking ways to easily access and share information with coworkers, customers, suppliers, educators, wherever these people may be. It's a new, much more powerful way of interacting, and it's evolving rapidly. By combining IBM and Lotus, we can make these benefits real for our customers more quickly.

Our intent is to keep Lotus intact and managed out of its present headquarters in Cambridge, Massachusetts. Lotus will also have primary responsibility for key, complementary IBM software products. We expect that Lotus's management and employees will remain and join forces with IBM's software people in creating a new model for enterprise computing.

IBM has communicated its offer to Lotus management. A copy of a letter to Lotus Chairman, President, and CEO Jim Manzi is attached. We are taking these actions after attempting on several occasions to open discussions that would result in a friendly merger.

The next step on our part will be to publish legal notices in tomorrow's *Wall Street Journal* and *The New York Times*.

I can't tell you much more at this point because of legal considerations, but I will keep you up-to-date as developments take place. In addition, a copy of this letter and other relevant material will be on the IBM home page on the Internet (http://www.ibm.com). In the meantime, if you receive any inquiries on our plans from outside the company, it is very important that you refer them to Corporate Communications.

At our annual meeting in April, I discussed the progress we have made in terms of productivity and cost-cutting. I also emphasized our intent to fully reestablish IBM's leadership position in the industry. We are well on our way to doing just that. A combination with Lotus will help us get there that much faster.

Louis V. Gerstner, Jr.

June 5, 1995

Mr. Jim P. Manzi
Chairman, President, and Chief Executive Officer
Lotus Development Corporation
55 Cambridge Parkway
Cambridge, Massachusetts 02142

Dear Jim:

As you know from your conversations with IBM Senior Vice President John M. Thompson, IBM has been interested for some time in pursuing a business combination with Lotus.

Because you have been unwilling to proceed with such a transaction, we are announcing this morning our intention to buy all of Lotus Development Corporation's outstanding common shares for a price of $60 per share, or $3.3 billion. This is an all-cash offer. We believe this is now the fastest, most efficient way to bring our companies together.

We have the highest respect for you and all Lotus employees. We believe our companies share similar visions of the future of information technology—a future built on a truly open, collaborative computing environment where people can work and communicate across enterprises and across corporate and national borders. Combining our efforts will mean that both of us reach that future sooner.

This is truly a win-win opportunity for IBM and Lotus shareholders, employees, and customers. With IBM's financial resources, technological expertise, and unmatched customer base, Lotus will have greater opportunities for growth and expansion.

With IBM's global marketing and sales capability, we can rapidly grow Notes' user base and vastly increase its sales potential and acceptance as an open industry standard. Working with industry partners and customers around the world, we will help them embrace this powerful new way of computing, working, and communicating. We also have the strength and resources to support Lotus's mail and application products.

Our objective is a transaction that is enthusiastically supported by you and the Lotus Board of Directors, as well as Lotus employees, shareholders, and your many loyal customers, software developers, and industry partners.

We respect the creative environment and entrepreneurial spirit you have fostered at Lotus. We do not want to change that. We believe Lotus's employees are among the best in the industry at developing innovative and successful products. Our intent is to keep Lotus intact and managed out of its current headquarters in Cambridge and to make Lotus primarily responsible for key, complementary IBM software products.

We and our advisors are prepared to meet with you and all other members of the Lotus Board of Directors, management, and advisors to answer any questions you or they may have about our offer. We are convinced that together we can achieve a business combination that serves the best interests of Lotus and IBM.

We believe, as you do, that the future of information technology is one in which anyone, anywhere will be able to share information and interact—easily and instantaneously—no matter where they are or what system they use. We look forward to working with you and your colleagues to develop products and systems that will allow customers around the world to realize this vision.

Louis V. Gerstner, Jr.
IBM Chairman and Chief Executive Officer

Appendix B

The Future of e-business

My original idea for this addendum was that it would be sufficient to collect a few of the industry keynote speeches I delivered during my time at IBM, reprint them as a chronology of how our e-business message evolved, and leave it at that.

Then I reread the speeches.

It was like staring into a large and unforgiving mirror. We got a lot of it right and accurately called more than our fair share of shots. But with the benefit of several years' distance from those speeches, it was just as clear that we missed a few predictions and trends, and some of what we thought would be important at the time turned out to be nothing more than the kind of experimentation and trial runs that are always the harbingers of true, sustained, technology-driven transformation. So be it. I'll stand by our record.

Instead of presenting a retrospective, I'll outline here how I anticipate e-business (and the progression of information technologies in general) will evolve. Following that, I'll present some observations on the potential impact on institutions, individuals, and all of society.

I have to ground this forward-looking discussion with a few statistical points of reference and acknowledge that three, four, or five years hence what will follow may seem more quaint than prescient.

There's a school of thought that says the world has a new mass medium when a technology is being used by at least 50 million people. Radio hit that threshold in about thirty years; it took television thirteen years; it took cable TV ten years. The Internet set a new standard. Less than five years after the birth of the World Wide Web, some 90 million people were connected.

By the summer of 2002 that number exceeded 500 million people. More than half of them were accessing the Web in languages other than English. While the estimates vary, several organizations that track this kind of thing said that worldwide Internet commerce would reach $4 trillion by 2005.

Without overstating the obvious, suffice it to say that the Net is more than a communications medium or a marketplace. Its exploitation is, and will be, the single most important driver of change in business, health care, government, education, and society. It is the transformational technology of our lifetimes, and that transformation is in the very early stages. I expect that the application of networking technologies will lead the agenda for at least another ten years before being replaced by biological sciences as the dominant technology in the world.

But let's remember that this amazing technology wasn't always seen that way, which, as I noted in Chapter 18, is what compelled us to create a new vocabulary around the term "e-business," in order to describe the broader, more powerful aspects of this change.

Like a lot of other world-altering technologies, the Net was ushered in amid a swirl of confusion and misinformation, and a very heavy concentration on what it was all going to mean for *individuals*. Perhaps you recall it: a celebration of the ultimate in the empowerment of every man, woman, and child with a Web browser accessing online magazines, downloading movies to their wristwatches, or buying pet food and flowers with a couple of clicks or keystrokes.

So, of course, when we at IBM said we believed there was something bigger happening than chat, browsing, or even online retail, a lot of people had a good time pointing out that, once again, plodding old IBM just didn't get it. And given the mood of those heady, early dot-com days, what we were saying *was* pretty boring.

We certainly agreed that the Net was going to change the world. But our perspective started with what was going to have to happen inside all of the world's existing *institutions*—banks, hospitals, universities, retailers, government agencies—to change the way they work, transform physical processes into digital processes, and extend their enterprises to the Net. Only then were *individuals* going to be able to do things—pay a bill, move money around, buy a stock, renew a driver's license—in fundamentally different ways.

Our message was essentially this: There is a new technology here that is going to transform every kind of enterprise and every kind of interaction. But please understand that this technology—like any other technology—is a tool. It is not a secret weapon or a panacea. It has not suspended the basics of marketplace economics or consumer behavior. And the winners will be found among the institutions that skip the shortcuts and understand that e-business is just business. It is about real, disciplined, serious work. And for those willing to do the unglamorous labor of transforming a process, unifying a supply chain, or building a knowledge-based corporate culture, it will deliver tangible and sustainable benefits.

In meetings that IBM hosted globally for hundreds of the world's leading CEOs, I liked to draw a comparison between e-business and the advent of electric power. Before people had the ability to generate electricity, a lot of what got transported in the world was moved by mules or horses. Then, over time, the activity of transporting things was taken over by electric-powered machinery. The industries didn't change. The basic activity of pulling and lifting didn't change. But the people who made the swiftest transition from the old technology (animals) to the new technology (machines) became the dominant players inside their industries. It was almost the same proposition with e-business.

That first experimental and speculative chapter in the evolution of e-business is fading from view. A second chapter—a far more serious and pragmatic period—is under way. Leaders in all industries see the benefits as well as the practical issues of implementation that they will face as they cross into the networked world, and they are seriously charting their individual strategic directions.

This coming phase of e-business will be characterized both by the technical implications, as well as a set of management and leadership challenges.

Bringing Down Barriers to Access

If you think about the proliferation of information technology, it took a remarkably brief period of time, less than forty years, for it to spread from the hands of a select number of centralized technicians—the high priests of the mainframe era—to tens of millions and then hundreds of millions of PC users.

The rise of the Net has made terms like "connected world" and "universal access" permanent additions to the lexicon of the twenty-first century. Yet the fact remains that more than half of the world's people have yet to make a phone call. The half billion Internet users I mentioned earlier is impressive for a technology that's still in its infancy, yet that represents less than 10 percent of the people on the planet. We're still a long way from the day when even a narrow majority of the world's population is firing up a browser and joining the community of people with access to the infrastructure of computing and communications. For the immediate future, then, we don't have to debate the existence of a digital divide between the world's information haves and have-nots. It's real. Its permanence, however, is another matter entirely.

Multiple factors contribute to this divide: disparities in education and literacy, telephone penetration, and access to electricity. In terms of computing and communications barriers, there are two:

telecommunications rates, and the cost of the access device itself. Both are coming down, one faster than the other.

Outside the G8 nations, world governments are taking steps to end monopolistic telecommunications practices, encourage competition, and open their markets to network operators and service providers. Most nations, nearly 80 percent, have opened their cellular markets to competition, though a majority of countries still retain monopolies (either state-run or privatized) in fixed-line services for local and long-distance service. Some citizens of the world can make a local three-minute call for 1 cent. Others pay fifty times that.

The second barrier—the expense of the access device itself—is rapidly being reduced. When the one and only access device was a full-blown personal computer, surfing the Web was an activity for the rich. But the world's entire inventory of hundreds of millions of PCs has already been eclipsed by an explosion of other kinds of low-cost access devices, from Net-enabled cell phones to personal digital assistants, game consoles, or even kiosks in marketplaces or government facilities. Within the next few years there will be billions of mobile devices (not counting personal computers) connected to the Net.

All of a sudden the price of entry is no longer an insurmountable barrier. Yet there are plenty of thoughtful people who assert that information technology will unavoidably and permanently separate the world into two camps: those with access, and those locked outside looking in. I do not accept the inevitability of their argument. It seems just as reasonable to me that with greater telecommunications competition, continued innovation by the IT industry, and thoughtful leadership at all levels of society, there is more than a chance—there is a magnificent opportunity—to shrink this gulf and spread unprecedented levels of service and information to people regardless of their social or political standing or personal buying power.

This proliferation of low-cost access devices is one dimension of the much more pervasive reach of information technology. But it doesn't stop there. Besides all the gizmos that people will actually use, the technology is literally vanishing into the fabric of our lives: the clothes we wear, the appliances in our homes, the cars we drive, and even the roads we travel—plus a thousand other things that we'd never think of as "computers." It's easy to envision a day when everything worth more than a few dollars will be outfitted with tiny chips, some storage, and communications capability. The applications are life-enriching, convenient, fun, practical, and powerful.

As just one example, when every product you own is continuously reporting its location, and "knows" whether or not it's where it's supposed to be, theft becomes a lot harder to pull off. For manufacturers and retailers, this all points to the next-generation in market analysis and customer service. Imagine the power of instantaneous information on every product they have in the marketplace—how it's being used, how much it's being used, and how it's performing. It's like getting Nielsen ratings on anything and everything—without the overnight wait. And for people and societies, consider the benefits of clothing that might warn the wearer of environmental hazards; of buildings whose architectures can adapt in the event of an earthquake; or water supplies capable of repelling attempts at sabotage.

It's all scientifically possible. When we'll see it, I wouldn't hazard a guess. Do I doubt that we *will* see it all, and more? Not for a second. Just look at what's already happened.

When I learned to drive, a car was a mode of transportation. Today some cars are a node on the Net. Many include a feature that reports the vehicle's location to emergency services anytime the air bags deploy. The world's leading manufacturer of pacemakers now equips them with Internet addresses that will one day contact your physician if anything starts to go wrong. If all kinds of appliances and heavy equipment sense that a part within them is failing, they can "phone home" to dispatch their own repair technician or to get the appropriate software download to fix the problem. IBM scientists are researching an "intelligent" kitchen counter, which would "read" medicine bottles placed side by side and issue a verbal warning if that combination of drugs could produce an adverse reaction. One Japanese company is even making pint-size beer glasses that would alert the bar staff when the mug is empty!

At each of these intersections—of the technology with devices, with people, and with the routines of everyday life—the role of the technology in our lives becomes more pervasive, and more invisible. It fades from view even as our expectations for what it can do increase.

On the other hand, what's happening behind the scenes to enable all these networked applications is dependent on a secure, global computing infrastructure. At that end of the computing continuum, things are taking on unprecedented levels of both sophistication and complexity. If we're going to keep moving forward—extending the reach and impact of the technology by making it easy to use—then masking its complexity becomes paramount.

Uncomplicating Computing

Enterprises of all kinds increasingly recognize the importance of entering the world of e-business. It's either that or consign themselves to the fate of those turn-of-the-century institutions that decided mule power suited them just fine. But as customers look down the road to digital nirvana, they see a road littered with potholes.

As we've seen, the ubiquity of computing becomes more real every day. Increasing numbers and kinds of devices are generating additional transactions, increasing data flows and network traffic, and all of it is happening with much greater unpredictability in usage and volumes. At the same time, threats to the security of systems and data are escalating far beyond what was predicted even a few years ago.

Leaders in the public and private sectors, in businesses large and small all over the world, know that e-business demands a fundamentally new kind of information infrastructure. It will be more secure, more capable, and more reliable than what is in place today. The dilemma (for them and for the people who make and sell the technology) is that the infrastructure has become almost impossible for customers to implement or manage.

The traditional remedy—throwing more people at the problem—simply won't work, not in the long term. Complexity is spiraling upward faster than the capability of humans to deal with it. Around the world, unfilled IT jobs already number in the hundreds of thousands, and demand is expected to increase more than 100 percent before the end of this decade. At this rate there simply won't be enough skilled people to keep the systems running.

Therefore, the infrastructure itself—from end to end—will have to be reengineered to have the ability to perform many tasks that require human intervention today. What is coming is a kind of computing that will take its cue from the human autonomic nervous system.

IBM's research scientists draw many parallels between the way the human body manages itself—everything from heartbeat to the immune system—and what is needed in computing systems. Think of it as a kind of self-awareness that will allow systems to defeat viruses, protect themselves from attack, isolate and repair failed components, see a breakdown coming and head it off, and reconfigure themselves on the fly to take full advantage of all of their component parts.

Autonomic computing won't be invented or created by one company alone. That's why IBM's technical community proposed in 2001 that this new realm would become the next great technical challenge for the entire IT industry.

Joining the Grid

So far the Internet and its communications protocols have enabled computing systems that were once self-standing—whether you're talking about PCs or data centers—to share information and conduct transactions. In effect, the first stage of the Internet revolution allowed computers to *talk* to one another. What's going to happen next (based on yet another set of gorpy protocols) will allow networks of computers actually to *work* with one another—to combine their processing power, storage capacities, and other resources to solve common problems.

This kind of massive, secure infrastructure of shared resources goes by the name "grid computing." Like many of the mainstream commercial aspects of information technology, such as the Internet itself, grids are taking off first in the scientific, engineering, and academic communities in areas such as high-energy physics, life sciences, and engineering design.

One of IBM's first grid projects was done with the University of Pennsylvania. It's designed to allow breast cancer researchers all over the world to collaborate on applications that will compare mammograms of the same woman over many years, leading to much more reliable detection and diagnosis.

The Next Utility

Put all of this together—the emergence of large-scale computing grids, the development of autonomic technologies that will allow these systems to be more self-managing, and the proliferation of computing devices into the very fabric of life and business—and it suggests one more major development in the history of the IT industry. This one will change the way IT companies take their products to market. It will change who they sell to and who the customer considers its "supplier." This development is what some have called "utility" computing.

The essential idea is that very soon enterprises will get their information technology in much the same way they get water or electric power. They don't now own a waterworks or power plant, and soon they'll no longer have to buy, house, and maintain any aspect of a traditional computing environment: The processing, the storage, the applications, the systems management, and the security will all be provided over the Net as a service—*on demand.*

The value proposition to customers is compelling: fewer assets; converting fixed costs to variable costs; access to unlimited computing resources on an as-needed basis; and the chance to shed the headaches of technology cycles, upgrades, maintenance, integration, and management.

Also, in a post-September 11, 2001, world in which there's much greater urgency about the security of information and systems, on-demand computing would provide access to an ultra-secure infrastructure and the ability to draw on systems that are dispersed—creating a new level of immunity from a natural disaster or an event that could wipe out a traditional, centralized data center.

Where will this take hold first? I think we're going to see something very similar to what we saw when customers started to embrace the Net. Many of the first implementations were for internal, or intranet, applications. In the case of on-demand computing, the ability to draw on a lot of existing resources plays directly to customers' questions about how to utilize fully all of their existing IT investments. Rather than rolling in another piece of hardware, buying a bigger database or more storage, customers could have a new way to leverage their existing resources.

The Outer Limits

It's almost always the case that any particular generation of people will be forced to deal with at least one game-changing technology—to understand it, apply it, and regulate it responsibly. In the middle of the last century, nuclear energy was the best example. The current generation, on the other hand, will deal with not one but two game-changing scientific developments. Everything I've described so far relates to the first—the implications for institutions and individuals associated with networking technologies. The second is what is happening around the marriage of information technologies with molecular biology.

The watershed event was the mapping of the human genome. That project created a data set equal to 10 million pages of information. Yet the really hard work is ahead of us. One researcher described it as having a book but not understanding the language in which it's written. Deciphering it is expected to require analysis of data sets at least 1,000 times larger than the mapping project itself—another 10 billion pages of information.

It will be worth the effort. What we'll learn will lead us to better, more effective, more personalized drugs, new protocols and treatments (and possibly cures) for the most intractable diseases, and new generations of more resilient, higher-yielding seeds and crops. The point is, there is huge potential here to limit human suffering and do with scourges like heart disease or AIDS what we've already done with polio and smallpox.

Eighty years ago, antibiotics ushered in the last great advance in the human life span—about twenty additional years over normal life expectancies in 1920. We're on the brink of discoveries that could deliver another twenty-year expansion, so younger readers of this book just might be looking at having a lot more time on planet Earth than their parents have had. And who wouldn't want it, since we're not talking about prolonging an existence already diminished by what we know today as "old age." We're talking about twenty more productive, healthy years.

The Real Issues Are Not Technical

Before we get carried away by what the technologies are making possible—at the networked level or at the cellular level—let's not forget that the potential societal good is always counterbalanced by an equally important list of societal concerns. And now that we have created the potential for both, it is my fervent hope that industry, customers, governments, and policy makers think through the implications of what is ahead.

It's already clear that a networked world raises many issues, such as the confidentiality of medical or financial records, or the freedom of expression v. protections of personal privacy. Think about the privacy implications of what's coming. What happens to personal privacy in a world of Internet-enabled cars that monitor our movements at all times; cell phones that continuously report their location; or Net-connected pacemakers and other medical devices that are gathering real-time data on our heartbeat or blood pressure, cholesterol level or blood-alcohol content? Who's going to have access to that most personal profile of you—your physician alone? Law enforcement agencies? An insurance provider? Your employer or a potential employer?

Earlier I mentioned the very real chasm that exists between the information haves and have-nots, and I expressed my hope that we might actually apply ourselves and these technologies to bridge that digital divide. As we do that work, however, I wonder if we're not in the process of creating a new, potentially unbridgeable genetic divide, where some people can afford the cost of preventing a birth defect or avoiding prolonged suffering, and some can't.

When advances in diagnosis and treatment converge to deliver on the promise of a longer, healthier life, have we merely created the priceless luxury of more time for the people and things we love? Or is there more to the equation? When that's possible—or well before it's possible—shouldn't we be thinking about the effect on social structures, the medical establishment, pension systems, and the environmental implications of having to produce more food and create more shelter?

Finally, after the events of September 11, 2001, we've all been forced to think about the greatest threats to our way of life, wherever we live in the world. Is it traditional military aggression? A rogue or state-sponsored terrorist attack? The danger of internal attack from a disenfranchised fringe element? There's no longer any need to say a lot about state-sponsored terrorism. We all view the world through a different lens now. One by-product of this new worldview is a basic rethinking of the nature of the threats we face—in all their forms.

Even after September 11 law enforcement and security agencies remained convinced that the greatest threat to people and societies was still posed not by weapons of mass destruction but by broad-based information warfare and what they call weapons of mass *effect*. No one equates the loss of human life with the loss of some computer equipment. At issue is the ability of cyber terrorists to cripple increasingly IT-intensive military infrastructures, national power grids, water supplies, or telecommunications systems.

The Leadership Challenge

This book has made the point repeatedly that leaders in both commercial endeavors and the public sector face a closely related set of strategic decisions about their exploitation of these technologies, their willingness to break with the status quo, their investment policies, and the readiness of their own leadership teams to embrace new ways of thinking and working.

That's front and center. Those choices are being made today. Tomorrow the agenda is going to shift to a set of considerations that revolve around what this networked world means for our existing geopolitical structures and all their underlying economic assumptions.

A networked world doesn't respect the fact that we've organized the world into nation-states and have adapted nearly every convention of life and society to that model. The course and development of a networked world is not governed by our concepts of national borders, regional alliances, or political structure. It's already dissolving many of the barriers that have historically separated peoples, nations, and cultures. And I believe it will drive a concomitant set of challenges to the ability of political institutions to control the most important thing they have always controlled—their citizens' access to information, education, and knowledge. In the process, we may see a shift in the way democracies behave.

How will governments arrive at workable policy frameworks in this globally, politically, and culturally connected world? On the issue of personal privacy, the European Union has a policy framework that's different from that of the United States, and both are markedly different from the Chinese approach.

Now take a step down from that level of global governance, to the way any individual anywhere in the world might express his or her political preferences. Not that long ago the thought of buying a book from your home or the office would have been considered revolutionary. So what happens if there comes a day when we can vote from the comfort of our den or the convenience of our workplace? Set aside what this might do to boost citizen participation in a representative form of government. Why not envision global referenda that are representative of a global populace voting without regard for political affiliations or national allegiances. What might it mean for individual governments when a world community expresses an opinion on issues like global warming or an agreement like GATT (General Agreement on Tariffs and Trade)?

I think that very soon, if we're not there already, there's going to be increasing conflict between what we would define as national interests and global interests. So we're going to be in a situation in which reaching agreement will require a new level of international cooperation and global public policy. But how?

Once again, our institutions are running well behind the rate of technological advancement. Some universities are starting to build e-business into their business management curricula. But what about political sciences, ethics, or the law schools? The United States Congress is one of the most powerful law-making bodies on the planet. To its credit there are a few committees and a handful of task forces examining issues like cyber-security, export controls, and intellectual property rights. But, for the most part, there's a fundamental lack of understanding about what it's going to take to build a workable policy framework for things like an appropriate tax regime for e-commerce.

In the second half of the twentieth century, the nations of the world came together to create multilateral institutions designed to foster economic growth, raise living standards, and forestall armed conflict. The United Nations, the Organization for Economic Cooperation and Development (OECD), the International Monetary Fund, and the World Bank are examples. I addressed the OECD in 1998. Much of my talk focused on the following questions: What is the global parallel of those organizations for the challenges of the Information Age? What global institutions do we need to create in order to play a similar stabilizing and enabling role in the twenty-first century?

All this leads me to consider whether we're looking at the requirement for what we might view as a new kind of leadership competency. It won't render obsolete the traits of successful leaders in the physical world. The Net is going to change many things, but not everything. Passion, confidence, and intelligence will always matter. As I've already noted in the discussion of the current crisis on confidence in business in general, integrity will matter more than ever before.

Yet I think it's typically the case that people who aren't forced to deal with the technology rarely make the effort to understand either its possibilities or its limitations. In the nuclear era, maybe that was all right. But for technologies as pervasive as the ones we're dealing with today, I believe we're going to need leaders in government, business, and policy-making roles who commit themselves to the challenge of lifelong learning in order to bring society into sync with the science.

This next generation of leaders—in both the public and private sectors—will have to expand its thinking around a set of economic, political, and social considerations. These leaders will be:

- Much more able to deal with the relentless, discontinuous change that this technology is creating.
- Much more global in outlook and practice.
- Much more able to strike an appropriate balance between the instinct for cultural preservation and the promise of regional or global cooperation.
- Much more able to embrace the fact that the world is moving to a model in which the "default" in every endeavor will be openness and integration, not isolation.

As someone who's just spent a decade inside the high-tech industry, I can say with confidence that its technologies are magnificent creations. But never believe that the technologies themselves come to us as self-contained answers. They are not mystical solutions to the most difficult and most important problems—like bias, poverty, intolerance, and fear—that have been with peoples and societies for all time. Those problems yield only to the most intensely human solutions—the kind that are devised by people of free will and self-determination, who possess the ability to choose and to decide, to think and to reason, and to apply the tools at their disposal to generate the greatest benefits, for the greatest number of people.

Appendix C

The charts in this Appendix summarize IBM's operational and financial performance for the years 1992–2001.

International Business Machines Corporation and Subsidiary Companies

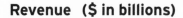

Revenue ($ in billions)

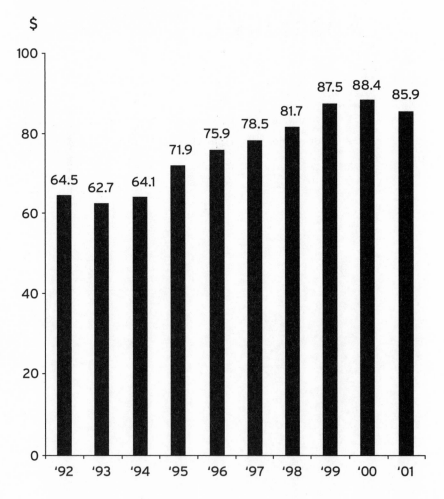

International Business Machines Corporation and Subsidiary Companies

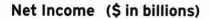

Net Income   ($ in billions)

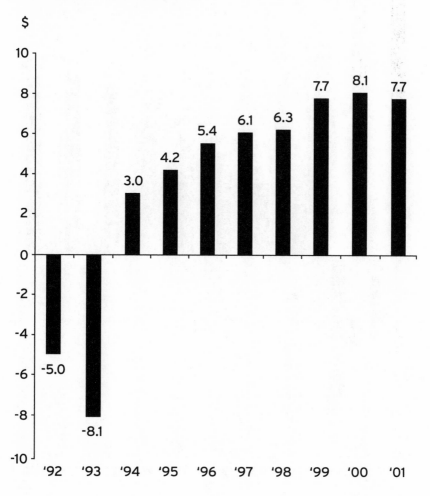

International Business Machines Corporation and Subsidiary Companies

Earnings per Share–Diluted

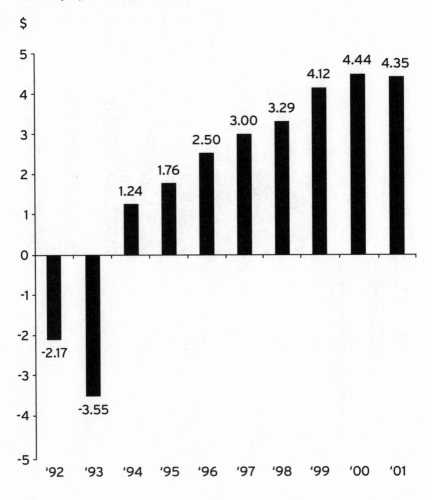

International Business Machines Corporation and Subsidiary Companies

Cash Flow from Operations   ($ in billions)

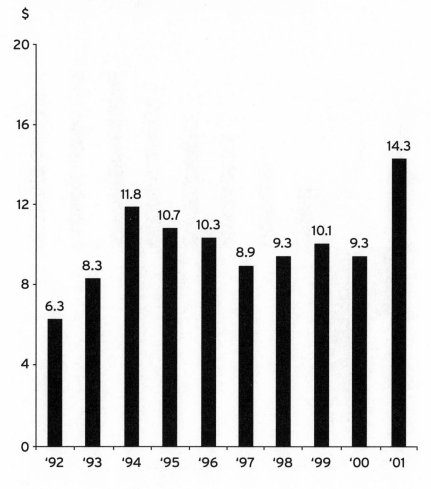

International Business Machines Corporation and Subsidiary Companies

Return on Stockholders' Equity

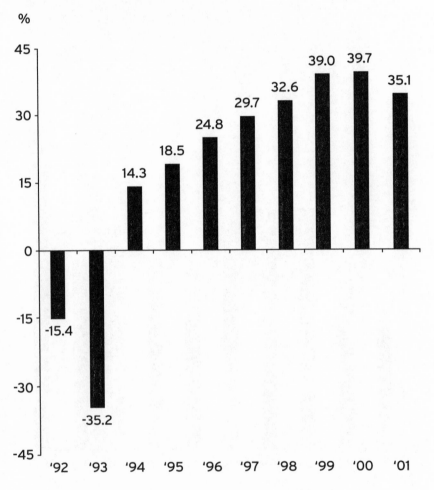

%

International Business Machines Corporation and Subsidiary Companies

Employees

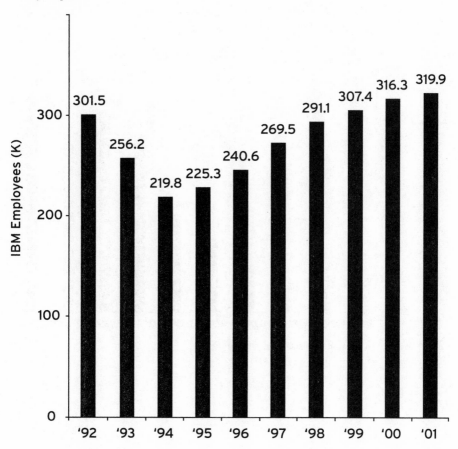

International Business Machines Corporation and Subsidiary Companies

Stock Price

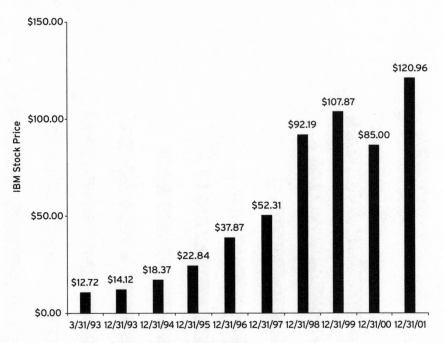

Adjusted for Stock Splits: May 1997, May 1999

International Business Machines Corporation and Subsidiary Companies

Revenue   ($ in billions)

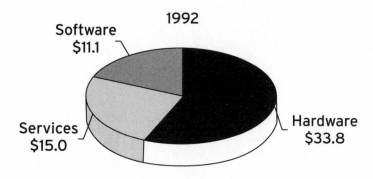

1992

Software
$11.1

Services
$15.0

Hardware
$33.8

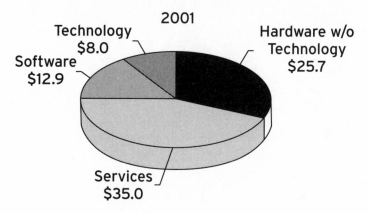

2001

Technology
$8.0

Software
$12.9

Hardware w/o
Technology
$25.7

Services
$35.0

Index